FETAL POSITIONS

WRITING SCIENCE

EDITORS Timothy Lenoir and Hans Ulrich Gumbrecht

FETAL
Positions

Individualism, Science,

Visuality

Karen Newman

STANFORD UNIVERSITY PRESS, STANFORD, CALIFORNIA

Stanford University Press
Stanford, California
© 1996 by the Board of Trustees of the
Leland Stanford Junior University
Printed in the United States of America

CIP data are at the end of the book

Stanford University Press publications are
distributed exclusively by Stanford University Press
within the United States, Canada, Mexico, and Central
America; they are distributed exclusively by Cambridge
University Press throughout the rest of the world

For Frances

ACKNOWLEDGMENTS

This project began, as most books do, while I was work-ing on its predecessor. Research for *Fashioning Femininity and En-glish Renaissance Drama* prompted me to read early modern anat-omies, midwifery books, and other medical texts. Subsequently, during five months spent in Bologna supported by a John Simon Guggenheim Memorial Fellowship, I visited the Museo oste-trico, about which I write in this essay. Without support from the foundation, I could never have begun this book. Seeing the Bolognese obstetrical models initially animated this project, but my theoretical interest in the presentism of cultural studies and the problems for contemporary politics, including the politics of the academy, posed by its turn away from history and "old stuff," determined me to pursue it.

I would like to thank my colleague Susan Bernstein for her generous friendship and astute reading of this essay at many stages of its composition; thanks as well for the skeptical commentary of the rest of an ongoing feminist reading group—Christina Crosby, Mary Ann Doane, Coppélia Kahn, and Ellen Rooney. Audiences at the University of Colorado, particularly Margaret Ferguson, Richard Halpern, and David Simpson; Dartmouth College; the English Institute; the Mellon lecture series spon-sored by the Department of the History of Art at Brown, the

Pembroke Center Seminar, 1991–92; and the Wood Institute for the History of Medicine at the College of Physicians of Philadelphia, provided lively discussion of this rather idiosyncratic project. Lisa Cartwright helped with bibliographic resources early on; Brooke Hammerle copied dozens of photographs, and Sarah Shun-lien Bynum helped with permissions, citations, and the dozens of mundane tasks it takes to get a book in print; despite his doubts about the body, Franco Fido shared his library; friends in Italy, Remo Ceserani, Anita Piemonti, and Giovanna Ceserani, helped with Italian materials—photographs, wildly expensive and beautiful books, references hard to find, and permissions hard to extract from the United States; and a conversation with Jonathan Goldberg driving at sunset back to Albuquerque from Santa Fe helped me find my title. It was my enormous good fortune to have Mary Poovey as one of the press's reviewers; she was the book's ideal reader. Dr. Santina Siena kindly helped with information about modern obstetrical training and practice. Errors are, of course, my own.

I have benefited from the collections of the University of Bologna and the Archivio di Stato in Bologna, the Specola in Florence, the Folger Shakespeare Library, the College of Physicians of Philadelphia, the John Hay Library at Brown, and the National Library of Medicine. Librarians at all these institutions were unstinting, but special thanks to Georgianna Ziegler at the Folger. I am grateful to Bryan Shepp, Dean of the Faculty at Brown, for financial help in assembling the figures and permissions, and to Ineke van Dongen for overseeing the process. Even so, without the generosity of the many people with proprietary rights to the images reproduced here, particularly Fabio Roversi-Monaco, *Rettore* of the University of Bologna, and Viviana Lanzarini, of the Museo Ostetrico, I could never have completed this book. Thanks as well to Helen Tartar for her initial interest in this project and her willingness to wait while it became shorter and shorter. Tom Brooks animates my life and work in untold ways; this book is written, as its dedication witnesses, for our daughter, Frances.

K.N.

CONTENTS

A NOTE ON FIGURES

In preparing the figures for this book, I have not always used images from the earliest edition of a particular text. Sometimes a later edition provided a better image for reproduction; sometimes obtaining a particular image from the earliest text was too difficult. In every case, however, I have checked the earliest available edition against the figure reproduced. In Sources and Credits, therefore, I have provided bibliographical information for the image reproduced and the date of the earliest edition consulted. Spelling of titles has not been modernized. In most cases I have cited the text from which an image is taken rather than the artist or engraver.

*B*ut the common form—with necks—was a proper figure, making our last bed like our first; nor much unlike the urns of our nativity, while we lay in the nether part of the earth, and inward vault of our microcosm.

Sir Thomas Browne, *Hydriotaphia*

*I*ndividualism, at first, only saps the virtues of public life; but in the long run it attacks and destroys all others and is at length absorbed in downright selfishness.

Alexis de Tocqueville, *Democracy in America*

FETAL POSITIONS

PROLOGUE

What follows is an extended essay about seeing science, about how science is presented in visual form and why it matters. Specifically it traces the early history of fetal images and the consequences of how obstetrical and embryological knowledge was represented over time in Europe, to both specialists and the public, as medical knowledge came to be produced and understood through anatomical observation. My argument is made as much via images as via verbal text, and it demonstrates Jacqueline Rose's observation that "there can be no work on the image, no challenge to its powers of illusion and address, which does not simultaneously challenge the fact of sexual difference."[1] Although my training is in the analysis of verbal texts from early modern Europe and England, here I consider a range of cultural and representational materials because scientific knowledge and the subjectivities it effects cannot be analyzed and confined to a given disciplinary field.

As I finished the manuscript for this book, John C. Salvi was arrested and charged with the murder of two workers and the wounding of five other people at two Brookline abortion and family planning clinics. On January 1, 1995, the year began with a lead story in the *New York Times*: "On the back of John Salvi's black pickup was an oversized picture of an aborted fetus. It was

a picture of Jesus Christ, Mr. Salvi told people."[2] The impact of
visual representation on the abortion debates calls for analysis;
this book argues that certain modes of visualizing science have
profoundly determined "fetal politics" and the contemporary
abortion debates. My purpose is to show the importance of history for framing and understanding contemporary political issues,
a point the often self-satisfied historicism of traditional disciplines
and the relentless presentism of cultural studies obscures. Recent
historicist work that might best be termed pseudo-Foucauldian
has taken the critique of historical cause and effect and of teleology, and the notion of Foucauldian epistemes, to mean that there
can be no diachronic arguments. Ironically, the claim on behalf
of a rupture, break, or shift separating one episteme from another
and foreclosing any diachronic possibility is teleology with a
vengeance. It presumes a *telos* reminiscent of those now de-
bunked Burckhardtian arguments for the break between the
Middle Ages and the Renaissance. The claim on behalf of dis-
crete epochs and a sovereign modernity attempts to absolve cultural critics from the hard work of understanding the past. As I
will point out, these obstetrical and embryological images differ
from one another in many respects—level of anatomical detail,
representational conventions, different media, and so on—but I
demonstrate that they share a core schema in the Gombrichian
sense.[3] The historical question is how that shared schema *means*
differently at different historical junctures: in the sixteenth century in relation to Aristotelian and Galenic notions of generation
and Albertian perspective; in the seventeenth century in relation
to the Cartesian *cogito* with its important links both to Renaissance perspective and to developing notions of subjectivity associated with the Enlightenment; in the eighteenth century in relation to political economy, an evolving public sphere, and the
production of a rights-bearing subject; in the nineteenth century
in relation to a positivist biologism; in the late twentieth century

in relation to the proliferation of rights claims, particularly feminism, the "technologization" of representation and reproduction, and changing notions of what constitutes "life." [4]

Finally, this essay is a contribution to the emerging study of the history of sexuality, in particular the necessary work of historicizing heterosexuality and reproductive practices. Recently cultural critics have argued that heterosexuality has a history and is not congruent with cross-sex relations in the past. But just as the advent of the homosexual in the late nineteenth century does not preclude work on queer practices and bodies and the anatomy of queer desire in earlier periods, so must work on early modern reproductive knowledge be part of any history of sexuality. [5] The history of reproduction and its relation to individualism that I analyze is material to the emergence of "modern" heterosexuality in the eighteenth century. [6] The study of gender and, recently, the focus on sexuality have rightly made us more precise in distinguishing gender from sexuality and both gender and sexuality from reproduction, but they are not unrelated categories. As Judith Butler has argued apropos race, gender, and sexuality, they are not "fully separable axes of power; the pluralist theoretical separation of these terms as 'categories' or indeed as 'positions' is itself based on exclusionary operations that attribute a false uniformity to them and that serve the regulatory aims of the liberal state." [7] My interest is in analyzing how the imaging of reproductive knowledge traverses different organizations of gender and sexuality.

In the opening pages of his recent and influential book *Making Sex*, Thomas Laqueur describes his experience as a medical student confronted with what is for him the undeniable distinction between the cultural body of history and the fleshly body of the dissecting table:

In my own life . . . the fraught chasm between representation and reality, seeing-as and seeing, remains. I spent 1980–81 in medical school

and studied what was *really* there as systematically as time and circum-
stance permitted. Body as cultural construct met body on the dissecting
table; more or less schematic anatomical illustrations—the most accurate modern science had to offer—rather hopelessly confronted the actual tangles of the human neck. For all my awareness of how deeply our
understanding of what we saw was historically contingent—the product
of institutional, political, and epistemological contingencies—the flesh
in its simplicity seemed always to shine through.[8]

In what follows, I argue that the human body as object of scientific study is, as the phrase goes, always already a cultural object
invested with meaning, "historically contingent—the product of
institutional, political, and epistemological contingencies." The
flesh, in fact, never simply shines through, as Laqueur would
have it, even in those "more or less schematic anatomical illustrations" that are "the most accurate" that curious creature "modern science" has to offer. Laqueur's italicized *really* allies him with
a realist position already adumbrated in his preceding binary, rep-
resentation/reality; similarly, his metaphor "to shine through"
assumes a dualist epistemology that opposes a "real" to some
sort of murky opacity that is its representation. The italics, however, should probably fall on the following adverb, *there*, since
the emphasis seems to be, "there is a there there," a posture that
for Laqueur would seem to require a retreat from the constructivist model to which he gives lip service ("body as cultural construct," "for all my awareness . . ."). But both contemporary literary theory and work in the social study of science have repeatedly demonstrated that the binary model real/representation,
with all its variants, is inadequate. A constructivist position need
not jettison the positing of a "real"; rather, it attempts to show
how that real—what realists sometimes refer to as "observ-
ables"—is saturated by all sorts of contingencies: perceptual,
temporal, historical, ideological.[9]

Cultural artifacts—anatomical illustration and models—link

the transient and situated work of the dissecting table to the relatively more lasting and extended world of "scientific literature" through which a scientific fact is produced and disseminated. I analyze a series of early modern envisionings of obstetrical and embryological knowledge and the move toward "naturalism" in medical representation they are said to demonstrate. My interest is in the rhetoric of early modern science and medicine rather than in specialized arguments peculiar to the history of science. My aim is to show how the visual can never be confined, as is sometimes claimed in medical illustration, to the "domain of simple recognition."[10] Since, quixotically perhaps, I want this book to be of interest to a wide audience, from those interested in sexuality and reproductive politics to specialists in the history of science, I have tried to confine its scholarly apparatus to the endnotes.

As the contemporary abortion debates witness, perhaps no flesh is more overdetermined with cultural meaning than the female reproductive body.[11] Commentators and activists on both sides of the abortion controversy have long recognized the importance of language and rhetoric in framing the debate.[12] Linguistic choices shape our attitudes: "pro-choice" versus "abortion," "anti-abortion" rather than "pro-life," "fetus" rather than "baby" or "unborn child," "uterus" rather than "womb." How *visual* modes of representing obstetrical and embryological information, which have similar consequences in forming both public and professional opinion, shape the politics of the abortion debates has until recently received very little attention. In the last decade, feminists have begun to analyze contemporary fetal imagery, but with little or no sense of the long and persistent history of such visualizations.[13] This book, with its interplay of visual and verbal texts, is designed to help the reader see the social work of consensus building that these representations achieve, in ways that their ubiquity and usual context inhibit. It analyzes the

principles of exclusion and inclusion, of schematization and elab-
oration, on which the visualization of obstetrical and embryological knowledge is based; it exposes the hierarchies and differences such images naturalize; it demonstrates the social and gendered roles they imply; it argues that these images inscribe a
certain sort of post-Enlightenment, rights-bearing subject crucial
to the contemporary abortion debates; and finally, it considers
briefly the possibilities and consequences of new medical technologies and postmodern visualities.

Fetal Positions

In a playfully ironic essay entitled "Visualization and Cognition: Thinking with Eyes and Hands," Bruno Latour assails the grand hypotheses of much contemporary theorizing about science and culture:

> Hypotheses about changes in the mind or human consciousness, in the structure of the brain, in social relations, in "mentalités," or in the economic infrastructure . . . are simply too grandiose. . . . No "new man" suddenly emerged in the sixteenth century. . . . The idea that a more rational mind or a more constraining scientific method emerged from darkness and chaos is too complicated a hypothesis.[1]

Taking his distance from contemporary materialist explanations, from Foucauldian epistemes, from Annales school mentalités, in fact from any sort of grand theory, Latour "deflates grandiose schemes and conceptual dichotomies" in favor of demonstrating how "groups of people argue with one another using paper, signs, prints and diagrams" (3). Instead of cognitive structures and paradigms, <u>Latour looks at science as "inscription,"</u> a term that embraces a variety of graphic modes for encoding scientific knowledge including maps, charts, prints and engravings, models, logs, reports, printed essays, and data banks. I use Latour's "inscription" rather than "visualization," the term perhaps more often used by those engaged in the social study of science, because it entails my continuing claim that there are no *visualizations* of science free of symbolic value, of social meaning.[2] Practices or "devices of inscription" are only significant because, in Latour's words, they "help to muster, align, and win over new and unexpected allies" (6). They are mobile; their symbolic value or social meaning can be marshaled on behalf of particular interests.

On January 22, 1992, at the annual march on Washington to protest the anniversary of *Roe v. Wade*, abortion opponents carried placards emblazoned with fetal photographs captioned "They're forgetting someone" (Fig. 1). So-called pro-life activists use such high-tech medical images on posters and billboards (Fig. 2), in pamphlets and advertising, and in film and video to persuade skeptics of their position. Feminist commentators have repeatedly observed that such fetal imagery effaces women's reproductive bodies. In feminist accounts, the image of a solitary fetus and the erasure of the woman's body demonstrate a certain set of social relations in which women and their bodies are subject to men—a paradigm more or less complicated by other categories, such as race, class, and/or historical context. Without in any way underestimating the importance of such ideological analyses of fetal imagery, I wish to account for <u>the extraordinary success the right has had in shaping how abortion is perceived</u> in the United States by following Latour's admonition. I will attend to inscriptions and how they are manipulated rather than hastening on to the grander ideological manifestos and conclusions that such inscriptions might imply.[3] As Latour puts it, "The weakest, by manipulating inscriptions of all sorts obsessively and exclusively, become the strongest" (32).

Fig. 1. Anti-abortion demonstrators, Washington, D.C. (*New York Times*, Jan. 23, 1992)

Fig. 2. (*Below*) Illinois billboard (late 1980s)

The fetal images of anti-abortion propaganda such as are re-produced in Figures 1 and 2 derive from the classic series by the Swedish photographer Lennart Nilsson, which has become the conventional mode of inscribing "fetal development" and the *lo-cus classicus* for feminist discussions of fetal imagery.[4] Through ex-treme enlargement and cropping to show only head and hands, the fetal image is manipulated to appear like a sleeping human baby. But these techniques to dramatize its human form are only the most obvious devices for inscribing fetal "life." Nilsson's se-ries first appeared in a 1965 *Life* magazine article entitled "Drama of Life Before Birth," almost a decade before *Roe v. Wade* and the subsequent anti-abortion movement.[5] Its deliberate linguistic strategies of persuasion—"baby," "person," "life," "womb"— and its mode of visual and rhetorical presentation helped to pro-duce the ideology of "fetal personhood" that has become the centerpiece of the "pro-life" movement.

As its title and opening paragraph establish, the "Drama of Life Before Birth" purports to represent "fetal life":

This is the first portrait ever made of a living embryo inside its mother's womb [Fig. 3]. It is one of an unprecedented set of color pho-tographs—strikingly complete in their clinical detail but at the same time strangely beautiful—of human embryos in their natural state. They were taken by Swedish photographer Lennart Nilsson, who worked seven years on his project. The embryos shown on the following pages had been surgically removed for a variety of medical reasons. But, using a specially built super wide-angle lens and a tiny flash beam at the end of a surgical scope, Nilsson was able to shoot this picture of a living 15-week-old embryo, its eyes still sealed shut, from only one inch away.

The paragraph is remarkable for its conspicuous ambiguities. It begins with the assertion, "This is the first portrait ever made of a living embryo inside its mother's womb." The word "portrait," defined in *Webster's* as "a painting or photograph etc. of a person, especially of his face," makes a claim from the outset for fetal

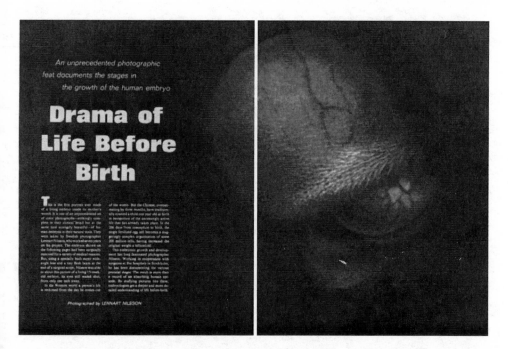

An unprecedented photographic
feat documents the stages in
the growth of the human embryo

Drama of Life Before Birth

Fig. 3. Fetus at fifteen weeks (*Life*, April 30, 1965)

personhood. Similarly, the choice of "mother's womb" is connotatively powerful in a quite different way from "uterus." Although *Life* represents the series of color photographs as "human embryos in their natural state," in fact only the initial photograph (Fig. 3) is of a "living embryo." The rest of the series, including the eighteen-week-old fetus that graces the cover (Fig. 4), were, as we learn several lines later, "surgically removed"—a technical detail of considerable importance.[6] Nilsson's uncropped photograph (Fig. 5) betrays what the *Life* cover conceals.

In short, Nilsson's photographs do not dramatize "life before birth."[7] They are photographs of fetuses obtained through both spontaneous and surgical abortion. Working in cooperation with

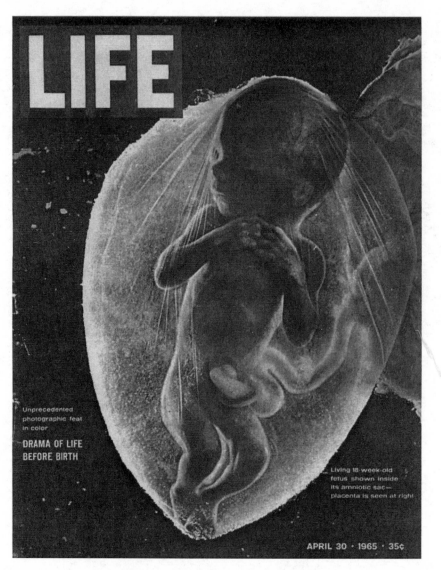

Fig. 4. Fetus at eighteen weeks (*Life*, April 30, 1965)

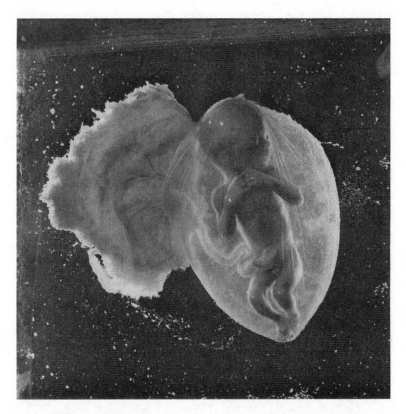

Fig. 5. Fetus at eighteen weeks (from Lennart Nilsson, *A Child Is Born*, 1965)

doctors in Sweden, where many privileged American women of sufficient means obtained abortions during the sixties, Nilsson perfected photographic techniques for chronicling embryonic development. Figure 6 shows an eleven-week fetus soon after miscarriage. By partially cutting away the placental mass, Nilsson revealed the two-and-one-half-inch fetus, which otherwise would not be visible to an observer; he then suspended the embryo in a clear fluid to facilitate the photographic process. Finally,

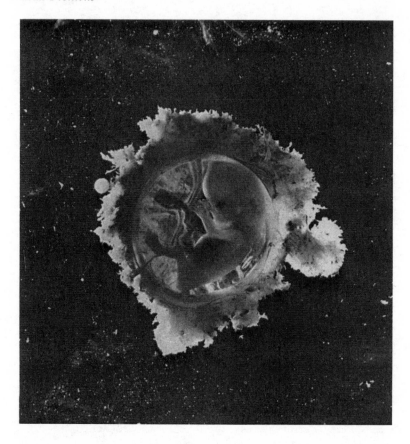

Fig. 6. Fetus at eleven weeks (Lennart Nilsson, 1963)

Life enlarged the fetal image to twice its actual size. Highly technical skills and complex instruments thus make "visible objects
and relationships which were invisible"—and which therefore
cannot be judged against a perceived real.[8] As the "Life Library
of Photography" volume *The Camera* describes it, "Whereas a
similar fetus preserved as a specimen in a laboratory bottle would
repel most beholders, Nilsson's painstaking technique has lent an

awesome beauty to this view of life at its beginning."[9] The "awesome beauty" of "life at its beginning" depends on a distinction between the grotesque medical specimen and an aesthetic of distance or voyeurism produced through instrumentation.[10]

In titling Nilsson's series "Drama of Life Before Birth," the *Life* editor uses "drama" in the sense of an unfolding series of events having unity and leading to a consummation—in this case, birth. But in its primary denotation, drama is a *composition* acted upon a stage in which a story is related by various means—gesture, costume, scenery. In that sense, then, the word also reminds us of the aesthetic artifice used to produce this "drama of life." Not only the lead paragraph but the entire *Life* essay is a tangle of contradictions as it negotiates between its claim to represent "life" and the various qualifications produced by the medico-technological processes, including backlighting, instrumental miniaturization and photographic enlargement, chemical and surgical preparation, and medical intervention, which are relegated to a series of parentheses and which continually admit of a quite different drama.

Whereas on the one hand properties and objects become observable/knowable through the graphic devices and practices used to represent them,[11] instrumentation is not enough; in addition, a normative rhetoric of apology is required to represent Nilsson's photographs to the *Life* reader. In the description of fertilization, for example, we are shown two photographs. One represents sperm, magnified 2,000 times, with no egg to fertilize: these sperm, we are told, "just mill around aimlessly." But when an egg is present, the sperm "stream purposefully toward it, as in the picture below" (Fig. 7). However, this description of the fertilization process ends with another series of parenthetical qualifiers—the photograph of the purposeful sperm was made "under laboratory conditions," a phrase followed by yet a further proviso: "(In the body only about 75% of the sperm would be

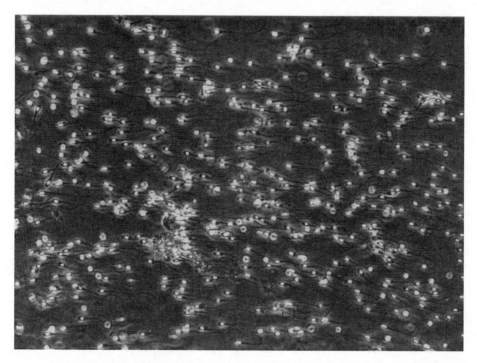

Fig. 7. Sperm (*Life*, April 30, 1965)

pointing in the same direction.)" Similarly in Figure 8, of a
three-and-a-half-week fetus, the caption reads: "This photograph is one of the few known to show face and head development at such an early stage"; subsequently, though, we learn that
in fact at five weeks "it is almost impossible to distinguish a human embryo from any other mammalian embryo"—seal, elephant, even rat.[12] Only a human embryo visualized via distancing
codes of scientific omniscience, isolated from the female uterus
and therefore cut off from any spatial identificatory cues, could
be confused with "any other mammalian embryo."[13]
     Over and over claims for representing "actual" fetal life and

the "living embryo" are made, only to be qualified: the placenta "has been partially peeled back"; "the amnion sac is wrinkled because some fluid has seeped out"; the "starlike spots around the amnion are merely bubbles in a fluid the photographer has used to support the amnion"; "the fetus has been backlighted." *Life*'s "drama of life before birth," as we learn intermittently, represents not life, but a simulation of life: in fact, life after death. And ironically, it is a death arrived at by the very means anti-abortionists oppose even as they deploy Nilsson's photographs on behalf of a putative fetal personhood.[14]

In Nilsson's later, portentously titled book *Behold Man: A Photographic Journey of Discovery Inside the Body* (1974), the human body is produced as a spectacle that pretends to the sacred.[15] A fetus at seven weeks is described as "a tiny human being"; at

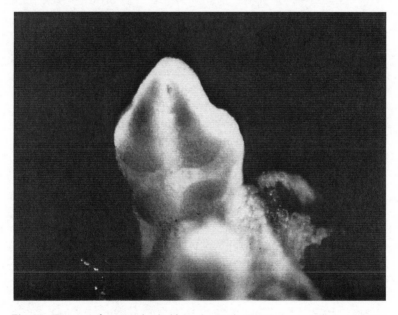

Fig. 8. Fetus at three and a half weeks (*Life*, April 30, 1965)

twelve weeks, it "moves freely in its capsule—weightless as an astronaut in space"; at sixteen weeks, it is "an active fellow."[16] The scriptural pretension of Nilsson's title and its representation of his photographic essay as new-world travel figures "man" as at once the subject and object of a colonizing gaze. The contribution of Nilsson's work to obstetrics and embryology has been challenged, but, notes one commentator, its "impact on public awareness of human ontogeny cannot be denied."[17] The technology of photographic inscription and its proleptic commentary produce the fetus as "living," as "natural," as a "human baby," "an astronaut," a baby who is rights-bearing, autonomous, and always already, in anti-abortion rhetoric, bourgeois.

At least since the Supreme Court's 1973 decision in *Roe v. Wade*, the debate over abortion in the United States has been framed in terms of rights.[18] Supporters of abortion speak of a woman's right to reproductive choice; opponents of abortion make claims on behalf of the putative rights of the unborn.[19] Implicit in rights claims are notions of individual autonomy and "personhood" that we associate with the Enlightenment. A quotation from Kristin Luker's *Abortion and the Politics of Motherhood* illustrates handily the intersection of liberal rights discourse, fetal personhood, and eighteenth-century political theory in the framing of the abortion controversy: "*The debate about abortion is a debate about personhood.* Whether the embryo is a fetus or a baby is important because virtually all of us agree that babies are persons and that persons have what our eighteenth-century ancestors called 'inalienable rights'—basic rights that cannot as a rule be lost, sold or given away."[20]

This liberal notion of rights has been appropriated by abortion opponents, as a proposed text to the so-called Human Life Amendment demonstrates: "The paramount right to life is vested in each human being from the moment of fertilization without regard to age, health, or conditions of dependency."[21] Anti-abortion activists explicitly recognize the significance of rhetoric in establishing personhood and rights claims. In his handbook *Closed: 99 Ways to Stop Abortion,* Joseph Scheidler, director of the Pro-Life Action League, instructs followers when speaking to the press: "Rarely use the word 'fetus.' Use 'baby' or 'unborn child.'" John Willke of the National Right to Life Committee exhorts followers to apply the "feminist credo" of "right to her own body" to the aborted fetus.[22] Demonstrators thrust plastic fetal models (Fig. 9) in the face of abortion clinic clients screaming, "Don't kill your baby," and fervently explain their mission: "We have to save the life of the pre-born human baby."[23] Slogans like "What about the baby's rights?" (Fig. 10) and "The baby has

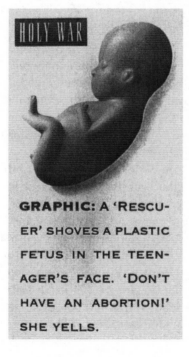

HOLY WAR

GRAPHIC: A 'RESCU-ER' SHOVES A PLASTIC FETUS IN THE TEEN-AGER'S FACE. 'DON'T HAVE AN ABORTION!' SHE YELLS.

Fig. 9. Fetal model (*New York Magazine,* April 24, 1989)

Fig. 10. Anti-abortion demonstrators, St. Paul, Minnesota (*New York Times*, July 25, 1993)

to have a choice"; Operation Rescue songs with lines like "Equal rights / Equal time / For the unborn children"; bumper stickers emblazoned "Equal rights for unborn women" or, in an ominous extension of anti-abortion rhetoric to birth control, "Every child has the right to be conceived"—all work to produce the fetus as a rights-bearing subject.

Following the murder on March 10, 1993, of Dr. David Gunn, who was shot in the back three times as he left his car to enter the Florida clinic where he worked, Don Treshman, national director of the anti-abortion group Rescue America, opined: "While Gunn's death is unfortunate, it's also true that quite a number of babies' lives will be saved."[24] A similar equation was set up by Robert Dornan, Republican representative from California, at a so-called Right-to-Life march in January 1991 when he regaled the public with his comparative view of the then-in-progress Gulf war and abortion: "Three babies are lost every minute [to abortion] and in six days we've only lost thirteen soldiers." [25] Dornan's comparison works through a series of conventional meanings presupposed in the juxtaposition of babies and soldiers. Soldiers, after all, are grown-ups; they train for combat and prepare for death. "Babies," however, are all the things babies are for our culture: innocent, small, helpless, endearing, imbued with possibility, in need of care and protection (Fig. 11). Despite a majority of voters who support choice, the right has co-opted the way abortion is framed in the United States: as a modern slaughter of the innocents. To persuade, anti-abortion activists use paraphernalia like white infant coffins filled with bloody baby dolls, strollers carrying mock skeletons, jars represented as containing aborted fetuses (Fig. 12), slide shows and film footage of intrauterine fetal movement, aborted embryos, severed fetal limbs and entire fetuses, even dead newborns. They refer to such visual materials as "war pictures" and describe them as "our most effective weapon" in what a *New York Magazine* article about Operation Rescue significantly dubbed a "Holy War." [26]

The language of warfare—killing, murder, weapons, strategies, tactics, Operation Rescue—and, recently, the escalation from language to physical violence against both property and persons serve to domesticate war by producing it safely within the boundaries of the nation-state. Faye Ginsburg describes the establishment of a permanent memorial dedicated to "all of our nation's children who have died by abortion" next to the war veterans' memorial in Fargo, North Dakota. Abortion, opponents claim, is war "on our unborn children. The veterans died to protect freedom everywhere, yet for the unborn there are no rights." [27] The analogy between babies and American war veterans, together with the disregard for the effects of war on those "other" peoples halfway round the world that is displayed in Dornan's analogy between abortion and U.S. aggression against Iraq, resituates the battle here at home. Public attention is turned

Fig. 11. Billboard, Providence, Rhode Island (1993)

Fig. 12. Abortion
opponents, Minneapolis
(*New York Times*,
July 25, 1993)

away from international conflicts, both those perpetrated by the United States and other cases of global aggression, to the "home front."[28] Such nationalist rhetoric might seem to evoke the social and a concern for the larger community, in contrast to individualism; it should not be forgotten, however, that the concept of nation is historically linked to the privileging of the individual.[29]

Abortion as an issue takes precedence not only over U.S. military aggression, but also over hunger, homelessness, neglect at home, and genocide and famine in the Third World. A layout in *Christianity Today* reporting on a Los Angeles rally to launch a cross-country anti-abortion protest juxtaposes a photograph of Melody Green, director of Americans Against Abortion, cradling an aborted fetus (Fig. 13) with a full-page advertisement on behalf of the starving children of Ethiopia (Fig. 14). The ad is illustrated with a photograph of one such child, its bloated stomach, spindly limbs, and hunched posture evocative of the exhibited, purportedly aborted, fetus opposite.[30] The article represents the displayed fetus as "the body of an aborted baby girl" that will "accompany the Walk America for Life team"—a palpable reminder that "aborted babies are more than just fetal tissue."[31] The emotional effect and the consequent political support for anti-abortion campaigns depend in large measure on the manipulation of visual images and a nationalist rhetoric that metamorphoses fetus into "baby" and leads to the rights claims entailed by that production of "personhood."

Feminists have begun to demonstrate how Enlightenment notions of individual autonomy and rights, notions that are central to U.S. constitutional law and liberal democracy, motivate and uphold both the argument for women's "right" to abortion *and* the concept of fetal personhood widely used by opponents of abortion. As Mary Poovey argues, "a metaphysics of substance" underpins rights discourse: "The basic assumption of the meta-

Fig. 13. (*Above*) Abortion opponent
(*Christianity Today,* July 12, 1985)
Fig. 14. (*Right*) Advertisement (*Christianity Today,* July 12, 1985)

physics of substance is that every subject has a substantive being
or 'core' that precedes social and linguistic coding. This substantive 'core,' which is the philosophical and putatively 'natural'
ground for legal 'personhood' and therefore for rights, is characterized by the capacity to reason, to exercise moral judgment,
and to acquire language."[32] The right-wing use of the language
of rights in many political arenas, from abortion to "free speech"
to the university curriculum, illustrates how highly problematic
and liable to appropriation is the discourse of rights and the notion of individualism on which it depends.[33] A vast literature has

already analyzed this Enlightenment discourse from the perspectives of modern political theory, philosophy, linguistics, and law. In what follows I want to consider instead Enlightenment modes of understanding and visualizing scientific knowledge about the body and reproduction, and the role that modes of visual inscription play in the production of the "individual" or rights-bearing subject on which both arguments for women's right to abortion and fetal personhood depend.[34]

Many commentators claim that the cult of fetal personhood is a phenomenon of the past thirty years. A 1993 essay in *Discourse* that analyzes Nilsson's *Life* photo essay, for example, claims that the "division between woman and fetus" is "historically unprecedented" and is the result of "a complex set of conjunctural circumstances," in particular new technologies of high-tech medicine and fetal visualization.[35] Roz Petchesky traces fetal imagery back to a 1962 issue of *Look* publicizing a then-new book, *The First Nine Months of Life*, that "featured the now-standard sequel of pictures at one day, one week, seven weeks, and so forth."[36] For Zoë Sofia, Stanley Kubrick's *2001: A Space Odyssey* (1968), with its footage of the fetal "star child" (Fig. 15), initiated the notion of fetal personhood.[37] There is no doubt that the media and new visual technologies have endowed the fetus with a public persona, a notoriety, even a star status, such as is epitomized in the following meditation:

A huge balloon containing a large-headed creature with four stumplike limbs floats near the Washington Monument during a rally addressed by [former] Vice President Dan Quayle. In 1990, this sky monster evokes the same object for every American: the fetus. In barely ten years, this disconnected figure has taken on a new symbolic character. . . . For little Mary, it now means the brother in her mother's belly. When it appears on a talk show, everyone knows that it stands for prenatal human life. When the law is at stake, it signals human rights. For some believing Christians, it is the smallest among the children of God and everyone's neighbor.[38]

Fig. 15. Still, *2001: A Space Odyssey* (1968)

However much a photograph's power of authentication may seem to exceed its powers of representation and thereby justify such ahistorical claims as Petchesky, Sofia, Duden, and others make,[39] in fact the presentation of the fetus as autonomous has a much longer history than these cultural analysts allow. That history is linked to anatomical illustration and sculpture, to conventions of classical representation that depend on an observing subject, to the development of modern obstetrics, and to Enlightenment science. If we are to understand the newly public status of the fetus, we must trace and analyze that history.[40]

Until the late seventeenth century, the visual codes for representing obstetrical knowledge were remarkably constant. The earliest illustrations appear in manuscripts of Muscio's Latin treatise, based in turn largely on a catechism of the second-century Ephesian physician Soranus's *Gynaecia* or *Gynecology*, in which some twelve to sixteen images illustrate various presentations of the fetus *in utero* (Fig. 16).[41] In these images, the uterus is rendered either in the shape of a jar or, alternatively, with earlike branches or horns that suggest adnexa and vasculature (Fig. 17); the fetal figures are likewise wholly stylized, conventional, and unrealistic homunculi, sometimes sticklike, sometimes pudgy, but always represented in fantastic gymnastic postures. These earliest visualizations of obstetrical knowledge illustrate a core schema that was reproduced well into the eighteenth century: a uterus separated from the female body and a seemingly autonomous fetal figure.

Similar images were adopted by Eucharius Rösslin for his *Der Swangern Frawen Rosengarten*, the best-known midwifery manual of the early modern period, which first appeared in German in 1513 and was subsequently translated into Latin, French, Spanish,

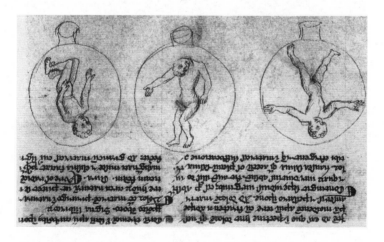

Fig. 16. From Muscio (13th cent.)

Italian, Polish, Czech, Dutch, and English. The book was frequently reissued into the eighteenth century, with some forty
editions published in English alone. *The Birth of Man-kinde; Otherwise Named the Woman's Booke*, as it was called in English, includes sixteen figures representing various fetal positions.[42] The
fetus is the same little man of the Muscio manuscripts, here

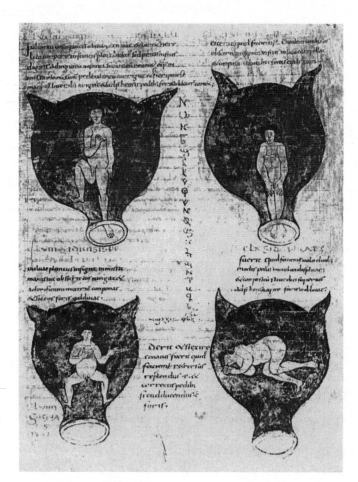

Fig. 17.
From Muscio
(9th cent.)

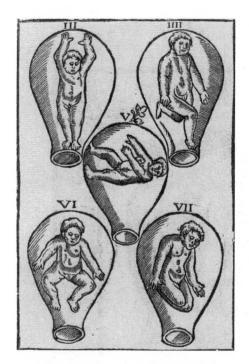

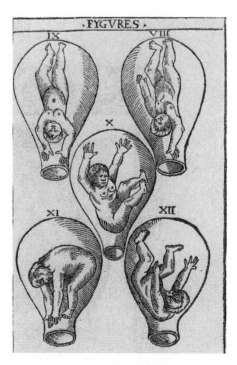

Fig. 18. From Eucharius Rösslin, *The Birth of Man-kinde; Otherwise Named the Woman's Booke* (1626)

Fig. 19. From Rösslin, *The Birth of Man-kinde*

plump and cherubic, jumping, dancing, diving, tumbling in unfettered freedom in a uterus represented diagrammatically, as in Muscio, like an urn or, anachronistically, a lightbulb (Figs. 18, 19).[43] In the course of the seventeenth century, although obstetrical and embryological knowledge became more complex, the basic schema for representing fetal positions persists. The Rösslin illustrations, which numerous midwifery manuals borrowed, acquired increasing detail, such as umbilical cords (Fig. 20), uterine layering, ovaries (Fig. 21), placental tissue, pelvic bone mass, and the like (Figs. 22, 23); nevertheless, the schema of the fully

Del modo di aiutare quel parto doppio, nel quale nascono due gemelli co' piedi auanti.
Cap. XIIII.

Fig. 20. From Scipione (Girolamo) Mercurio, *La comare o riccoglitrice* (1618)

Fig. 21. From Jacob Rueff, *The Expert Midwife* (1637)

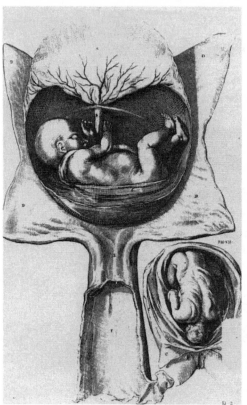

Fig. 22. From Hieronymus
Fabricius, *De formato foetu* (1627)

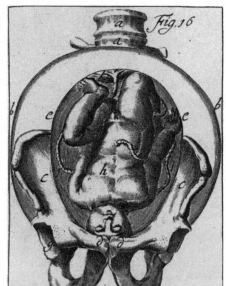

Fig. 23. From Hendrik van Deventer, *Operationes chirurgicae novum lumen exhibentes obstetricantibus* (1701)

formed fetus actively negotiating the uterine environment and cut off from a female body endures (Figs. 24–27).

The entire series of disembodied wombs from the Soranian tradition through the many editions of Rösslin and into the handbooks of the late seventeenth century suppress completely fetal dependence on the female body by graphically rendering that body as a passive receptacle, the scriptural woman as "vessel"—a visual rendering in keeping with the medical belief in "preformation," in which the fetus was conceived of as preformed, a fully fashioned though tiny adult that simply grew in size.[44] In early medicine, the uterus was believed to be passive and the fetus active during labor, with birth taking place thanks to the autonomous efforts of the fetus, conceived of as a "small hero breaking his chains, overcoming the bonds and restrictive or oppressive forces of his womb-world."[45]

These early images present an ideology of gender—the stereotypically passive female body-as-vessel and a conversely active, always-represented-as-male fetus—as "scientific" knowledge. Even after it was established that birth took place by means of uterine movement, the belief in fetal autonomy persisted, modified such that now the fetus was said to stimulate the uterine activity leading to delivery. In other words, both scientific discourse and its graphic encoding produced profoundly gendered stereotypes as "knowledge."[46] Even in those early midwifery texts that attempt to show stages of fetal growth, the fetus is a tiny homunculus lost in the vast expanse of the uterus, connected to an umbilical cord reminiscent of the mythological Ariadne's labyrinthine thread (Fig. 28). Not until the late eighteenth century do images appear that suggest the size and significance of the uterus and placenta or fetal dependence on the woman's body (Figs. 29, 30).

Leonardo's late fifteenth-century drawing of a fetus *in utero* (Fig. 31) is always cited in opposition to the Soranian fetal images. The artist is said to be the first to have rendered the fetus in

34

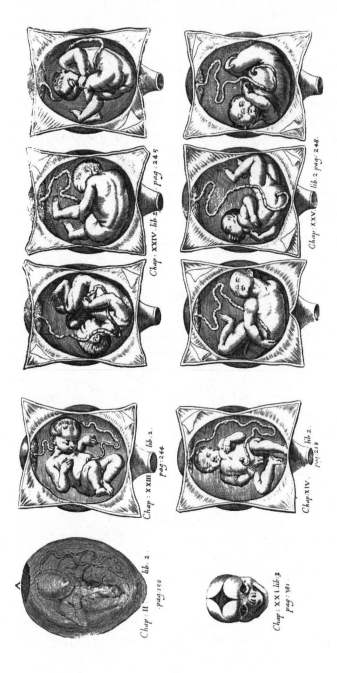

Fig. 24. Francis Mauriceau, *The Diseases of Women with Child and in Child-bed* (1683)

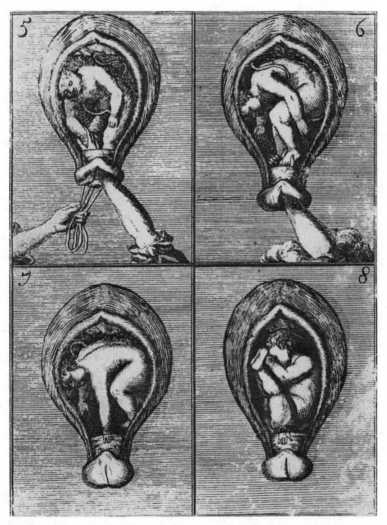

Fig. 25. From Justine Dittrich Siegemund, *Spiegel der vroed-vrouwen* (1691)

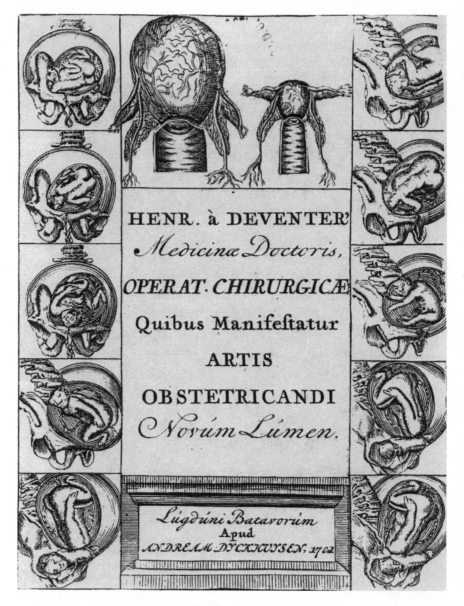

Fig. 26. From van Deventer, *Operationes chirurgicae*

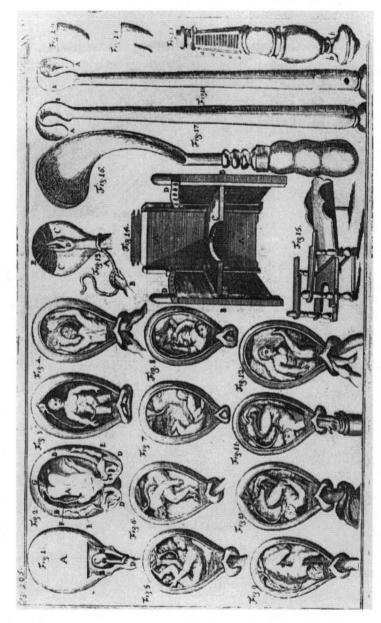

Fig. 27. From Lorenz Heister, *Institutiones chirurgicae* (1740)

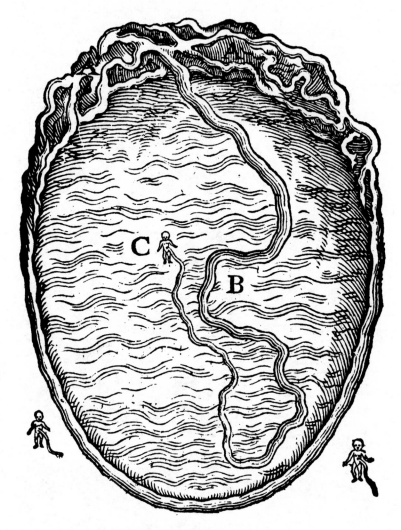

Fig. 28. From Severin Pineau, *Opusculum physiologum et anatomicum* (1597)

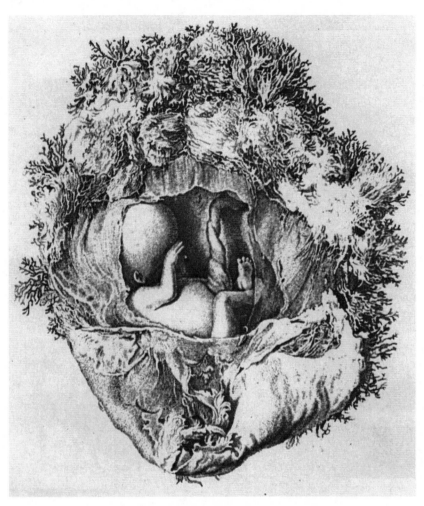

Fig. 29. "An Human Ovum, about the third month" (1783)

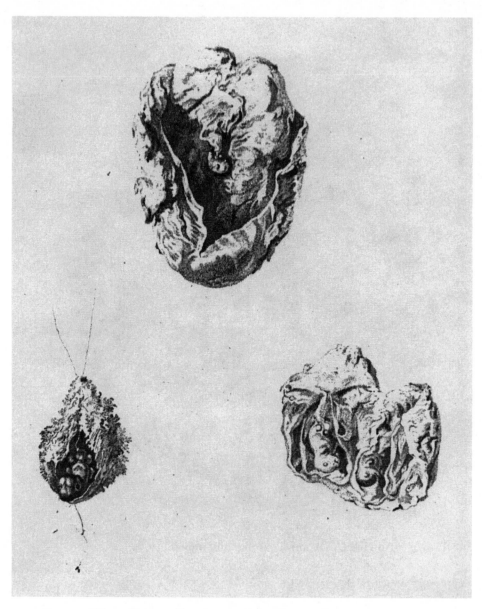

Fig. 30. "Three Human Abortions 23 February 1787"

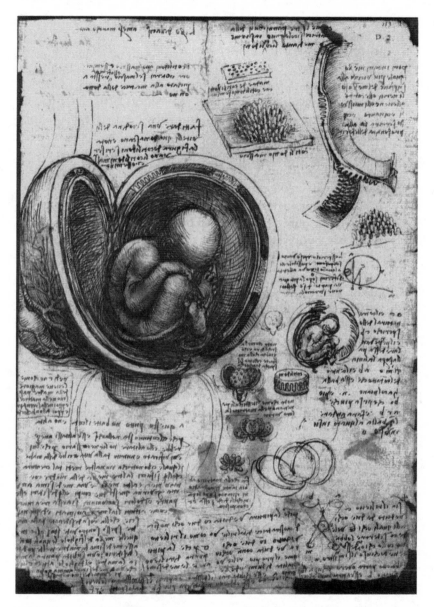

Fig. 31. Rendering by Leonardo da Vinci (after 1487)

its proper aspect in the uterus—in what we now refer to as the "fetal position." Certainly in comparison with the roughly contemporaneous Rösslin woodcuts, Leonardo's sketch is remarkable for its naturalistic detail. It and others of his life drawings have inspired generations of commentators to see in his work all the platitudes associated with the idea of "the Renaissance" as the rebirth of Man in all his individuality.[47] Yet Leonardo's design signifies beyond any canons of the realistic or natural. Instead of soft tissue and clinging folds, the uterus remains a vessel, but here it recalls a spherical box—almost a Fabergé egg—that opens to reveal its treasure, the curled-up baby within. The umbilical cord discreetly outlines the curve of the fetal thigh and buttocks, but there is no indication of dependence on the maternal body. In Leonardo's drawing, there is certainly a proliferation of realistic detail, but it is far from "realistic" for all its *effet de réel*.

The art historian Norman Bryson argues that "realist visual regimes . . . typically prioritise visual cues in a descending order, from the *definitive* attribute, to secondary qualification, through to the semantically innocent detail."[48] In theorizing the recognition and understanding of images under "realism," Bryson argues for a system of denotation and connotation in which, perhaps paradoxically, an "excess of connotation over denotation . . . constitutes the effect of the real" (67). Interpretation, or what Bryson terms "hermeneutic effort," is required to decipher images rich in connotative detail (64). As an example, Bryson traces how meaning is produced by formal oppositions in Giotto's representation of Judas and Christ (Fig. 32). The viewer is called upon to interpret the two figures according to features that are "subject to no absolute test of correctness, and this absence of univocal criteria confirms that the provenance of the information is *interpretation*" rather than denotative truth (64). The effort to extract meaning therefore takes place "under conditions of difficulty and uncertainty," which produce the sense of meaning "as *found*, not made; . . . meaning is felt to inhere in an objective

Fig. 32. Detail from
Giotto, *The Betrayal of
Christ* (ca. 1304–13)

world and is not apprehended as a *product* of particular cultural
work" (64).

Scientific illustration would seem to complicate Bryson's epistemological schema for realism, since it often eschews both
iconographic meaning and the formal oppositions characteristic
of painting. In anatomical representation, for instance, there
would always seem to be an "absolute test of correctness." But in
fact we find the same superfluity of detail in anatomical illustration as in representational painting, such that "connotation so
confirms and substantiates denotation that the latter appears to
rise to a level of truth" (62). If on the one hand, then, these obstetrical images seem to present a quasi-objective medical vision
tending toward the diagrammatic and nonindividuated—toward
what Bryson terms the "deliverance of the body's *typicality*"
(127)—on the other they provide an almost unbounded oppor-

tunity for the transgression of their own generality through the proliferation of excessive detail. In the course of the seventeenth century, details are added to the core schema of uterus-fetus— first to the uterus, but increasingly from the mid–seventeenth century on to the fetus, which is endowed with seemingly "gratuitous" detail that exceeds the instrumental function of the obstetrical image. The addition of seemingly innocent minutiae in fetal representation works to render the fetus as "lifelike" and "human," an autonomous individual.

To read diachronically by tracing the conventional mode of imaging obstetrical knowledge back over time is not to read ahistorically or to collapse history. Although we can see the fully formed fetus actively negotiating the uterine environment and cut off from the female body in Soranus, Muscio, Rösslin, Mercurio, Mauriceau, and Heister, these images had quite different meanings at their respective historical moments. The medieval and early Renaissance images, for example, represented the commonplace Galenic notion that proper pregnancies eventuate in male births. The suppression of the female gestating body acknowledged belief in the formative male role at conception, a notion crucial to Aristotelian generation. The images may even incorporate assumptions about the Christian soul. My interest here, however, is in tracing how these representations were used and interpreted in the late seventeenth century during the formation of the Enlightenment, and in the eighteenth century when they proliferated. Such representations were one of the many discursive modes that contributed to Enlightenment conceptions of individualism.[49]

ℱ

Obstetrical knowledge was inscribed not only in book illustration but, from the early seventeenth century on, in anatomical sculpture as well.[50] Perhaps the most extraordinary example is

Bologna's obstetrical museum, one of a series of small "pedagogical" museums at the University of Bologna. Although obstetrical models from the eighteenth century are also preserved in Padua, Florence, Modena, Pistoia, and Rome, all are based on the Bolognese collection and were usually fabricated in that city. Bologna's Museo ostetrico was the first and only collection to present the entire range of obstetrical knowledge and practice; it consists of an astonishing collection of early obstetrical "furnishings" and instruments, including some 130-odd models of uteruses and fetuses. All are arranged in glass cases along the walls of a good-sized room on the second floor of the Palazzo Poggi, the main administrative building of the university.

Today the approach to the collection takes the visitor first through a large, formal meeting room, hung with memorabilia of the university's long history, including a kind of secular last supper in stone that represents in conventional fashion one of the university's well-known *professori* at a long table imparting his wisdom to young scholars, represented as comically miniature to emphasize their master's superior intellectual stature. One then usually passes through another of the university's small museums, a military collection of mock fortifications in hardwood, topographical models, and displays of such memorabilia as Mussolini's university report card. Adjacent on the left is a tiny naval museum, filled with ship's models used to teach design and boatbuilding, its walls hung with early maps that chronicle exploration and colonization. Having passed through the "how-to" of war on land and sea, the visitor arrives at last at the Museo ostetrico, which represents not the strategies of warfare, but those of birth. Nothing escapes the logic of musealization.[51]

The patron of this extraordinary collection was Giovanni Antonio Galli (1708–82), represented in Figure 33 with the iconographic tool of his trade, surgical forceps, then newly in use. Galli was the first professor of obstetrics at the Istituto delle

Fig. 33. Giovanni
Antonio Galli
(1708–82)

Scienze and at the university, and in his time was one of the most
renowned medical men in Bologna. A doctor and surgeon, he
practiced and taught surgery in the city from 1736 until his
death.[52] He was thus very much a part of the scientific community of his era, influenced by Enlightenment protocols of science
based on collections, models, and morcellement, empirical observation and investigation, with "findings" regularly published
in the prestigious journal *Commentarii*.[53]

During the 1740s—also the moment of the crafting and assembling of the great wax anatomical collections in Bologna and
subsequently at the Specola in Florence—Galli began commissioning and assembling this obstetrical collection,[54] which he displayed in his home and used in the instruction of local midwives

(said to be the first such school). The earliest models were made in wax by Giovanni Manzolini (1700–1755) and Anna Morandi Manzolini (1716–74), the husband-and-wife team (Figs. 34, 35) that executed the extraordinary anatomical sculptures for the university's anatomical museum; but the expense of works in wax for what was, after all, the training of presumably illiterate women soon led Galli to commission his uterine models to be made from common clay.[55] In 1757, the entire collection was purchased by Pope Benedetto XIV, who assigned it to the Istituto delle Scienze. Within the year it was established in the Palazzo Poggi, and a course in obstetrics was instituted, comprising some sixty lessons annually.

Fig. 34. Giovanni Manzolini (1700–1755)   Fig. 35. Anna Morandi Manzolini (1716–74)

Fig. 36. Frontispiece
of S. W. Fores, *Man-
Midwifery Dissected*
(1793)

The collection, though assembled to teach midwives, became increasingly the province of male doctors, a shift demonstrated in local records that describe a *portella*, or little door, installed as a special entrance for women practitioners so they could enter "without their having to come in by the main door and gain entry [*introdursi*] into the Institute."[56] *Introdursi* means to be introduced into society or into a circle or group, and figuratively, to penetrate or intrude; its use here reveals the stakes of educating midwives together with aspiring doctors. What is now the Museo ostetrico was thus founded at an important transitional moment in the history of European medicine, when birth began to pass out of the hands of women and into the instruments of men, a transition graphically rendered in Figure 36, an English satiric print called "A Man-mid-wife."[57]

Interest in the practice and management of obstetrics—in "scientific midwifery"—expanded in the late seventeenth and eighteenth centuries as political economy became linked to population through theories of mercantilism. State power and expansion, economic prosperity, and military security were all considered to be dependent on demographic growth. These doctrines and their consequent policies fostered new discourses of public health that led to hospital expansion, the regulation of midwifery, and the professionalization of obstetrics.[58]

Until the late seventeenth century in Europe, normal births among the non-elite seem to have been handled mostly by women (Fig. 37).[59] Midwifery was a craft, passed on from mother

Fig. 37. From Ovid, *Trasformazioni* (1553)

Fig. 38. Twins (wax; 1740s)

to daughter or other female kin through apprenticeship and practice.[60] Doctors, by contrast, were certified in medicine and philosophy and were often sharply separated from the bodies of the sick, particularly those of women.[61] Although the regulation of midwifery in Europe dates from the late 1400s, by the mid–eighteenth century women began to be required to have some formal training, lasting anywhere from a few months to two years, in schools associated with either a university or a hospital. In northern Italy and Tuscany, for example, some thirty midwifery manuals aimed at professional standardization were published, many addressed exclusively to women; in some locales, however, women were simply prohibited from practicing midwifery.[62] As obstetrics became increasingly technologized with the invention and manufacture of forceps and ever more elaborate pelvimeters and specula, and as an accompanying increasingly positivistic view of anatomy produced the body as a series of parts to be manipulated, obstetrics became professionalized and, ultimately, the province of men, especially among urban elites. Professionalization also sparked a concern for public health and initiated a discourse about infant mortality that often blamed women practitioners, lobbied for their education, regulation, or prohibition, and promoted the male professional.[63]

The Bolognese obstetrical museum's 130-odd uterine and fetal models are arranged as they were in Galli's time, for pedagogic purposes, according to an inventory prepared in 1757 when the collection was sold and transferred to the Istituto delle Scienze. The first fourteen models are wax and include an extraordinary example of twins (Fig. 38), as well as a series of mounted pelvises. The next series, nos. 15–23, all in terra-cotta, represent the uterus at monthly stages of fetal development (Fig. 39). Only in no. 24 (Fig. 40) is the uterus placed within the "body," but in an abbreviated torso, truncated, mounted on wood, and draped, its relation to the whole body repressed. The abdomen is slit open,

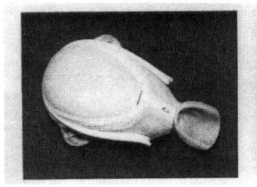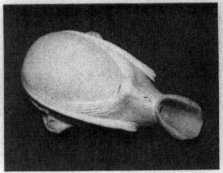

Fig. 39. Pregnant uterus at monthly stages (1740s)

Fig. 40. (*Facing page*)
Pregnant uterus at
seven months, in
opened torso (1740s)

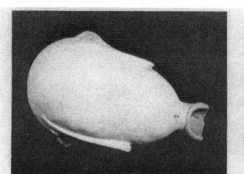
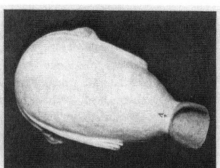
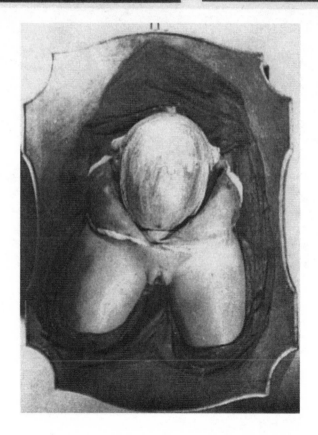

the layers of skin and muscle folded back, the pubis hairless, the thighs mere stubs.

Next there are various views of the placenta, models showing the uterus in various stages as birth approaches, and images of the fetus's imagined movements at birth.[64] Several examples of the birth sac and placenta follow, like giant eggs set on turned wooden pedestals (Fig. 41). Only one model, used to teach "nor-

Fig. 42. Removal of the afterbirth (1740s)

mal" deliveries, permits manipulation of the fetus inside the uterus; others depict the removal of the afterbirth (Figs. 42, 43). The remaining 86 models illustrate the problems of birthing that can result, for example, from "inexpert hands" (Fig. 44), from a difficult presentation of the fetus (Figs. 45, 46), or from trouble presented by the angle of the birth canal. There are a few examples of postpartum conditions, including prolapsed uteruses

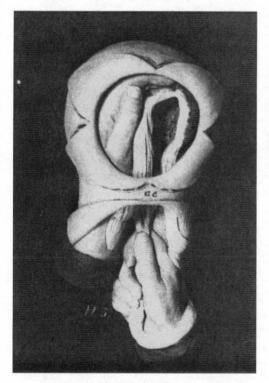

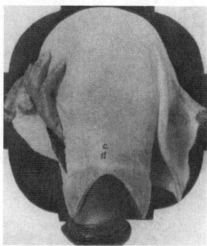

Fig. 44. Pregnant uterus lacerated during attempted delivery (1740s)

Fig. 43. Removal of the afterbirth (1740s)

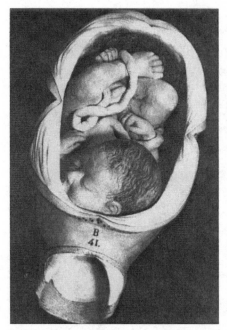
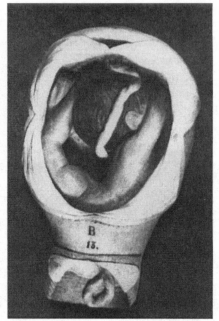

Fig. 45. Difficult presentation (1740s)   Fig. 46. Difficult presentation (1740s)

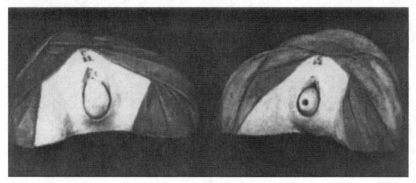

Fig. 47. Prolapsed uterus (1740s)

protruding from vaginas, their hairless pudenda represented as if independent of the body (Figs. 47, 48). Two uterine models, apparently together with a fetal doll, were used in pedagogic demonstrations (Figs. 49, 50), and several models of monstrous births, believed to exemplify the power of the maternal imagina-

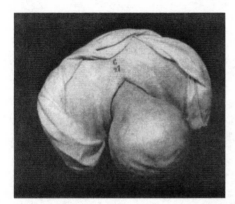

Fig. 48. Prolapsed uterus (1740s)

Fig. 49. Uterine model in wire (18th cent.)

tion to deform the fetus—a topic on which Galli often lectured at the university—are included as well (Figs. 51, 52).[65]

A birthing chair and numerous early obstetrical instruments (Figs. 53–55), including examples of the speculum, the *tiratesta* used to remove a stillborn fetus, and forceps, complete the collection. In both the museum catalog and its video counterpart, the mode of photographic representation of the obstetrical in-

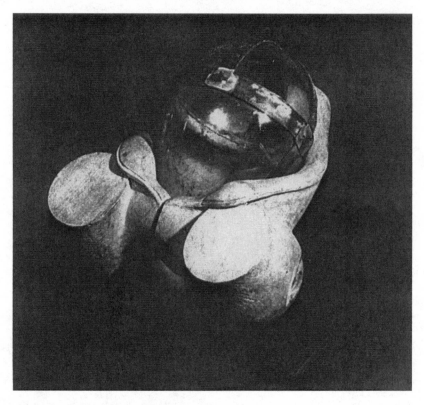

Fig. 50. Uterine model in crystal (18th cent.)

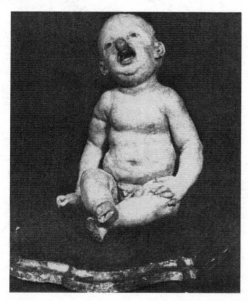

Fig. 51. Monstrous birth
(18th cent.)

Fig. 52. Monstrous birth
(18th cent.)

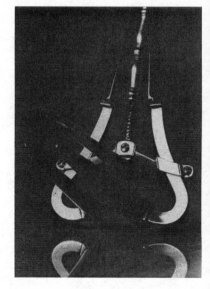

Fig. 53. (*Right*)
Speculum
(18th cent.)

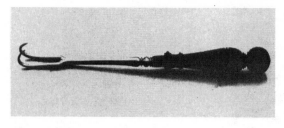

Fig. 54. Obstetrical hook (18th cent.)

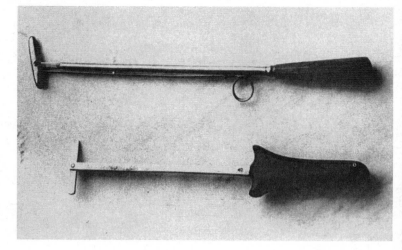

Fig. 55.
Tirateste, or
head pullers
(18th cent.)

struments, particularly forceps (Fig. 56), detaches them from their utilitarian purpose, lighting them dramatically so as to aestheticize and thereby suppress the frequently horrific damage that these instruments could inflict on fetus and mother alike. The photographs also emphasize the phallic shape of the instruments, which were produced by a male-dominated technology linked to the growing status of obstetrics.[66]

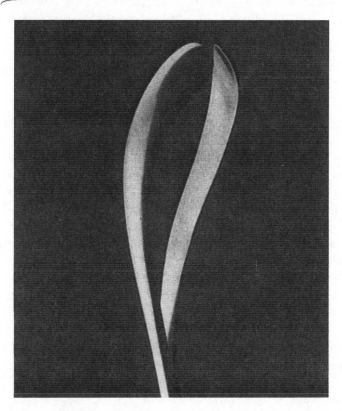

Fig. 56. Forceps (18th cent.)

The uterine models, with their wealth of detail, construct gestation and birth in ways that at first seem startlingly naturalistic, in striking contrast to earlier obstetrical images. In the Bolognese models, each uterus contains a fetus with sculpted hair, tiny curled hands, eyes sealed shut, in appropriate fetal positions. Instead of having the enlarged head, scrawny limbs, downy skin, and squinting eyes of the full-term fetus or neonate, these fetuses look like babies two, even three months old—plump, hirsute, with filled-out cheeks, in peaceful slumber (Fig. 57). Enveloped in their clay wombs, they recall, as Marco Bortolotti observes in his introduction to the museum catalog, the many sculpted baby Jesuses that were found in crèches all over eighteenth-century Bologna. They thus produce the fetus as a living infant, enlisting the conventions of contemporary religious sculpture with its powerfully emotive effects on behalf of a newly scientized obstetrics.

The realistic effect of these models depends not only on how the fetuses appear discretely or singly, but also on the series.[67] The characteristic mode of representing fetal development and the stages of birth serially and syntagmatically inserts each sculpted fetus in a "narrative" of developing "life." This view is dramatized by the museum's video catalog, in which a clay uterus is shown alone on a pale background, filling virtually the entire frame; with each shot fading to the next, superimposition produces the illusion of fetal growth until, miraculously and without labor, a baby is born.[68]

This narrative of reproduction adheres to an epistemology invested in visibility and to a reproductive politics that is situated in a particular mode of seeing. This mode of seeing, in turn, is produced by hegemonic conventions of classical representation that continue to regulate how we interpret fetal images.[69] Renaissance perspective takes vision as the basis for representation; more specifically, it adopts a particular point of view, that of the unified, seeing subject:

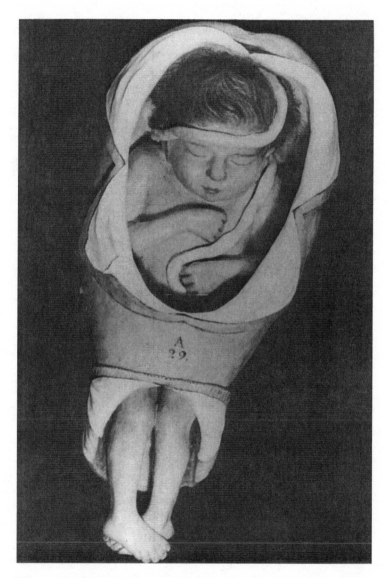

Fig. 57. Fetal presentation (1740s)

The vanishing point acts as a mirror, reflecting back to the spectator an imagined version of himself. . . . Each image within the code of perspectival art thus offers the spectator the possibility of objectifying himself, the means of perceiving himself, from the outside, as a unitary seeing subject, since each image makes a deictic declaration: this is how I see (or would see) some real or imagined scene from this particular spot at this particular instant in time. Being able to signify such a particularized individuality equips the perspective code with the visual equivalent of a demonstrative pronoun, allowing the code to deal in messages whose interpretation requires the active presence of a physically located, corporeal individual who has a "point of view."[70]

This regime of visibility facilitated discourses of power over numerous classes of objects and bodies. With regard to the human body, the art historian Michael Baxandall in his work on Renaissance sculpture theorizes that "the disposition to infer character and feeling from a representation of a human figure is both strong and deep. . . . We are all very skilled in interpreting visual appearance. . . . Posture, gesture, glance, the fixed lineaments of the body and the face. In particular we are sensitive to what all these imply of an attitude toward ourselves."[71] Recognition of the lifelikeness of the engraved or sculpted human form depends on "ourselves," as Baxandall puts it—on an observing subject produced by the epistemological position of perspectivalism linked to the Cartesian *cogito*.[72]

Early modern obstetrical visualizations—woodcuts, engravings, and sculpture—figure the female body not merely as inert, an object, but, in its openness, its breached boundaries, and its capacity to "make babies," as a violation of what Donna Haraway has termed "liberal singularity."[73] At the same time, these images encourage identification with the whole fetal body, rendering it seemingly transcendent and autonomous. This visual narrative of reproduction and selfhood inscribes a certain Enlightenment subject that, in Hal Foster's words, "subtends metaphysical thought, empirical science, and capitalist logic all at once" while at the same time displaying the woman's incapacity as an "individual" (from *individuus*, indivisible).[74] In his useful entry in *Keywords*, Raymond Williams points out the contradictory meanings of the word "individual" and alludes to the significance of its shift in meaning for political theory, economics, and scientific thought: "*Individual* originally meant indivisible. That now sounds like paradox. 'Individual' stresses a distinction from others; 'indivisible' a necessary connection. The development of the modern meaning from the original meaning is a record in language of an extraordinary social and political history."[75]

What we now term individual rights and powers ground modern political theory; likewise, obstetrical illustration and sculpture encourage what Baxandall calls "the disposition to infer character," a psychic identification that helped produce what C. B. Macpherson has labeled "possessive individualism."[76] These obstetrical images recall Locke's quotation of scripture in his chapter "Of Property" in the *Second Treatise of Government*: the earth is given, he says, "to the children of men." For Locke, private property is not a social institution, but rests in the individual's autonomy: "Every man has a property in his own person; this nobody has any right to but himself." Such "natural" autonomy and self-sufficiency are exemplified by reproduction in the chapter called "Of Paternal Power," in which Locke claims that "every man's children, being by nature as free as himself or any of his ancestors ever were, may . . . choose what society they will."[77] Locke's emphasis on the children of men, and particularly sons, allows him to address the social reproduction of the individual as citizen without evoking the woman's body whence it issues. As Haraway has argued,

Pregnant women in western cultures are in a much more shocking relation than men to doctrines of unencumbered property in the self. In "making babies," female bodies violate western women's liberal singularity during their lifetimes and compromise their full claims to citizenship. . . . Ontologically always potentially pregnant, women are both more limited in themselves, with a body that betrays their individuality, and limiting to men's fantastic self-reproductive projects.[78]

Woman's failed singularity—her reproductive body—justified the refusal to extend to women the rights claimed for universal "Man" and thereby helped to deflect the threat to gender hierarchies posed by Enlightenment liberalism.

Theories of preformationism were in fact the scientific counterpart and underpinning of political theories of individualism. In discussing early embryology in her work on the eighteenth-century man-midwife William Hunter, Ludmilla Jordanova points out that "preformationism . . . removes the need to think about foetal growth and the human status of the child [*sic*]. Within this framework, the child was always human and complete; it merely had to grow bigger."[79] Even after epigenesis—the notion that the fetus changes over time not merely in size, but by developing organs, limbs, and so forth—became accepted following the publication in 1759 of Kaspar Friedrich Wolff's *Theoria generationis*, belief in "the human status of the child" persisted.

Current medical dictionaries in the United States, as Paula Treichler observes, often still ignore the woman's labor in childbirth and present birth as a feat performed by the fetus: it is defined as "the act or process of being born" and "the emergence of a new individual from the body of its parent."[80] The new medical specialty of "fetology" depends on a similar inscription of the fetus as active, independent, individual.[81] Commentators claim that fetal personhood has come about only with the advent of new visual technologies that render

the once opaque womb transparent, stripping the veil of mystery from the dark inner sanctum. . . . The sonographic voyeur, spying on the unwary fetus, finds him or her a surprisingly active little creature, and not at all the passive parasite we had imagined.

The fetus has come a long way—from biblical "seed" and mystical "homunculus" to an individual with medical problems that can be diagnosed and treated, that is, a patient. Although he cannot make an appointment and seldom even complains, this patient will at times need a physician.[82]

Likewise, the editors of a collection entitled *The Unborn Patient* dedicate their book to "Gretchen, Antoinette, and Barbara" (presumably their wives) "and to the courageous mothers who, by pursuing treatment for their unborn babies, enfranchised the fetal patient."[83] The autonomous human child represented in midwifery manuals, obstetrical atlases, anatomical sculpture, modern medical dictionaries, and new medical specialties is independent of the woman's body, whole and undivided, always male, and virtually never dissected, opened, wounded, or permeable; it is the image *par excellence* of rights-bearing Enlightenment Man ferociously rendered in the fabled state of nature.[84]

Feminist commentators point out, of course, that fetal imagery is far from realistic in its representation of obstetrical knowledge. No attempt is made, for example, to represent the fluids surrounding birth or the musculature that enables the woman's labor. Even in naturalistically executed models, such as those in Bologna, the uterus remains the same passive vessel of the earliest obstetrical illustrations. The combination of representing the fetus as a fully formed child and suppressing the connection of the reproductive organs to the woman's body, many feminists claim, elicits sympathy for the "baby" and inhibits emotional response to the "mother."[85] The composition of obstetrical images—a complete, undissected fetal body and a schematic, or even invisible, uterus that conceals fetal dependence on the female body, serving instead as mere setting—constructs a narrative of reproduction in which the fetal figure is central, its context marginal. In Roz Petchesky's words, "The foetus is solitary, dangling in the air (or in its sac) with nothing to connect it to any life support system. . . . From their beginning such photographs have represented the foetus as primary and autonomous, the woman as absent and peripheral."[86] This isolation of the uterus from the female body, it is claimed, emphasizes fetal personhood and erases the already fully human status of the woman giving birth.

But reinsertion of the "missing mother" in the imaging of birth to include the woman's pregnant body, and presumably thereby to recognize her role in and significance to the process of birth, simply reinscribes what might be termed the cultural logic of individualism. The woman *in absentia* that feminist commentators lament is, after all, just one more liberal rights-bearing subject. Those who react against the reductionism of fetal imagery and would reclaim for representation the woman's body forget that the "individual" was in fact the initial phase of the reductionist project. As Richard Lewontin points out, "The Cartesian commitment to reduction that was meant to justify the replacement of the collective by the individual as the locus of action annihilates the individual as the locus of action, annihilates the individual on its march toward the quark."[87] Restoration of the woman's body to fetal images produces the woman herself as individual: another disenfranchised subject with rights who must be represented, another figure of humanist identification and sympathy.

૪

And presence, after all, is no guarantee. Anatomical illustration and sculpture, unlike obstetrical visualizations, frequently represent the full female form or torso, with a fully formed fetal body inside, but without conferring "personhood." From Andreas Vesalius's remarkable engravings for the *De humani corporis fabrica* (1543) to the extraordinary wax models of the human body, its organs, skeleton, and musculature, produced in Italy in the mid-eighteenth century (and still to be seen not only in Bologna, but also in Florence, Vienna, London, and Philadelphia), the scientific information of anatomy is rendered via the canons of conventional artistic representation.[88]

Much has been written on the traditions of both anatomical illustration and wax sculpture; I will therefore confine my discussion to issues pertinent to my larger argument. It has often been noted that in anatomical representation, at least since Vesalius, the various physiological systems—vascular, muscular, nervous, lymphatic, and so forth—are typically sexed (and sometimes gendered) male. A male body, that is, represents the norm. Even in that great eighteenth-century tribute to rationalism, the *Encyclopédie*, among 33 plates that illustrate the article "Anatomie," no female body is represented. Female anatomy is represented on a single page containing an illustration of the uterus; another illustration of the placenta with a fetus standing independently alongside, the umbilical cord reaching not directly to its opened torso but, oddly, behind the head and over the shoulder; and three details of the vulva that focus on the clitoris (Fig. 58).

Medical knowledge is visualized in early anatomy in and on a male body; female anatomy, it seems, can only be represented synecdochically in its sexual/reproductive specificity. The opened female torso presents the uterus and other reproductive organs and virtually always includes a fetus, represented as a fully formed tiny adult, a homunculus (Fig. 59). The female body in which the fetus is lodged is rendered in a variety of ways—as a classical nude (Figs. 60, 61), as a courtesan (Figs. 62–64), as a *venus pudica* (Figs. 65, 66), or as Eve before the fall (Fig. 67)—but always in her reproductive capacity. This anatomical tradition is also related to religious iconography, in which Jesus appears as a fully formed miniature adult inside the Virgin's womb (Figs. 68, 69).[89]

Even later, ostensibly more sophisticated seventeenth-century anatomical texts such as those of Spieghel and Bartholin render

70

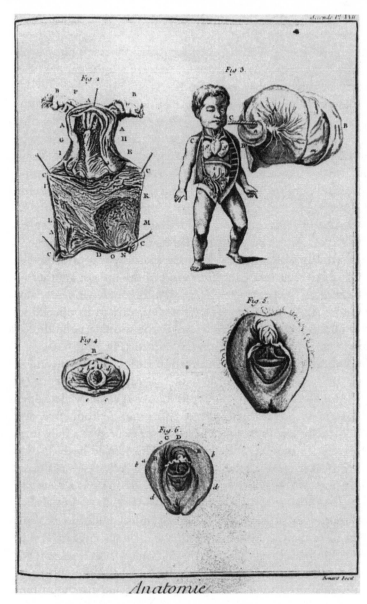

Fig. 58. "Anatomie," plate XXII, *L'Encyclopédie* (1762)

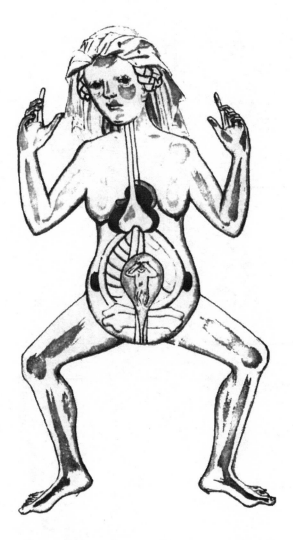

Fig. 59. From a fourteenth-century manuscript

72

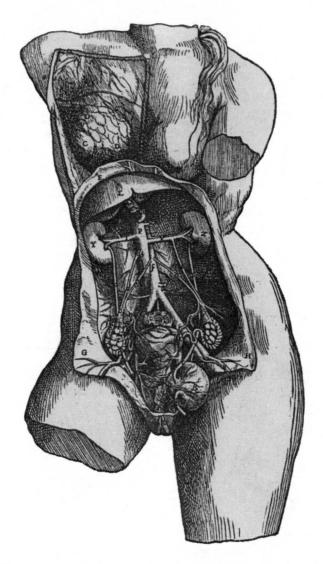

Fig. 60. From Andreas Vesalius, *De humani corporis fabrica* (1543)

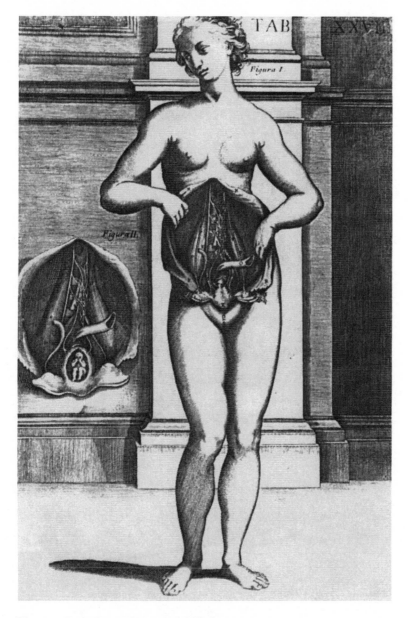

Fig. 61. From Petro Berrettini, *Tabula anatomicae* (1741)

Fig. 62. From Cosme Viardel, *Observations sur la pratique des acouchemens naturels* (1673)

A Secundina disse-
Eta, vsque ad al-
lantoidem.
B Facies secundi-
næ, ad allantoi-
dem peruenien-
tes.

Fig. 63. From Charles Estienne, *De dissectione partium corporis humani* (1545)

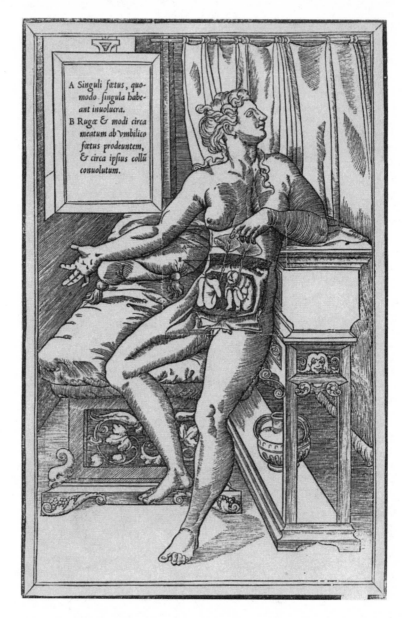

A Singuli fœtus, quo-
modo singula habe-
ant inuolucra.
B Rugæ & modi circa
meatum ab vmbilico
fœtus prodeuntem,
& circa ipsius collũ
conuolutum.

Fig. 64. From Estienne, *De dissectione partium corporis humani*

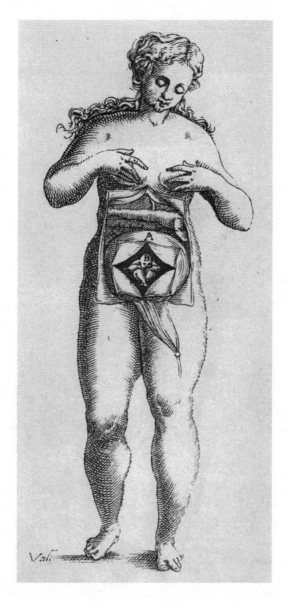

Fig. 65. From Scipione (Girolamo) Mercurio,
La comare o riccoglitrice (1601)

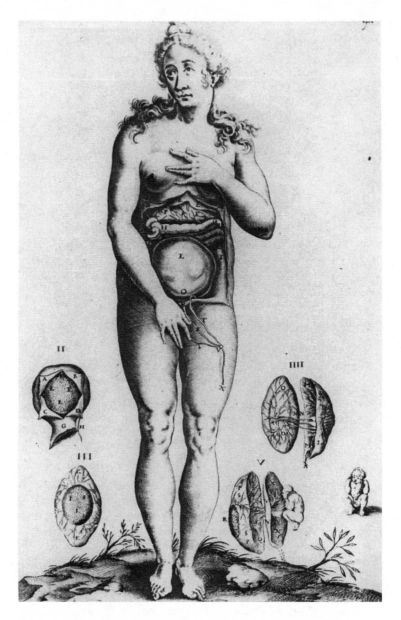

Fig. 66. From André Du Laurens, *L'anatomie universelle* (1731)

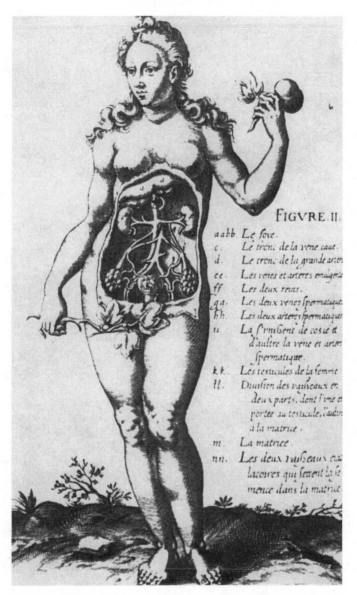

Fig. 67. From Jourdain Guibelet, *Trois discours philosophiques* (1603)

Fig. 68. Visitation of Mary and Elizabeth, Schotten Altarpiece (1390)

Fig. 69. Mary (1720–30)

female anatomy via overdetermined gender codes. Spieghel's female body is rendered as Eve, an apple partly visible behind her back, her sexual parts modestly hidden by fortuitous flora—apparently the iconographic lily, not as yet in bloom, of the angel Gabriel and the Virgin at the annunciation, thus encoding at once Eve's sin and the sign of human redemption (Fig. 70). Her body itself is pictured as a flower, its petals layers of skin and muscle. The physiological details of those anatomical layers are expunged to emphasize abstract aesthetic effect. In Bartholin's version (Fig. 71), the maternal body is a torso with petal-like layers and merely ornamental vasculature that frame an infant in beatific slumber, a conventional rendering, as Figure 72 likewise illustrates. In other words, the female body is appropriated into a series of gendered narratives—mythological, biblical, classical humanist, anthropological (woman as nature/flower)—codes that discipline the cultural threat posed by her reproductive body and assure the self-evidence of sexual difference. The woman-as-reproductive-body, as mother, cannot be allowed to encroach upon or trouble the identificatory relay between observing subject and fetal body. The privilege of disembodiment that defines the individual paradoxically requires the profoundly embodied and coded bodies of women, but those bodies must always be represented at some remove from the fetus and the process of birth.[90] Not until the end of the eighteenth century are there representations of a woman's fully pregnant body outside such coded narratives (Fig. 73). Significantly, the illustrated uterus and fetus accompanying the image are also strikingly different: the uterus and placenta are rendered with more detail and are sized proportionately, and the fetus, at an early stage of development, is marked by unformed features and limbs.

In the great eighteenth-century wax anatomical collections, the cadaver, that fixture of medical study, is a supine woman in

Fig. 70. From Adriaan Spieghel,
De formato foetu (1626)

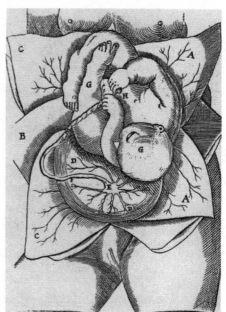

Fig. 71. From Thomas Bartholin,
Bartholin's Anatomy (1668)

Fig. 72. Frontispiece of Hendrik van Roonhuyze, *Heel-konstige Aenmerckingen* (1663)

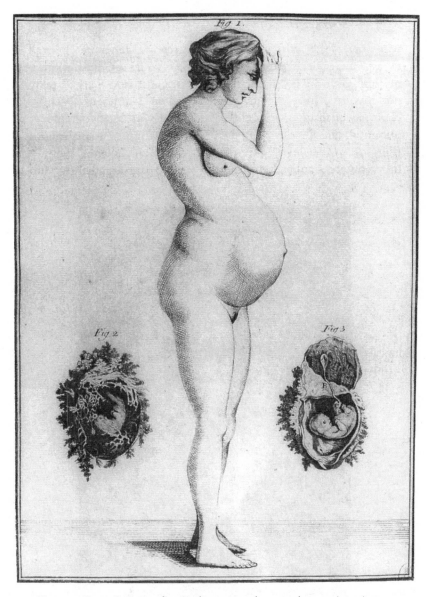

Fig. 73. From J. B. Jacobs, *Ecole pratique des accouchemens* (1793)

wax (Fig. 74).[91] Although these figures are generally referred to as "anatomical venuses," the earlier codes of illustration—mythological, biblical, anthropological, and so on—are notably not invoked. Instead, they are inserted in a different semantic chain, as odalisques, sometimes bedecked with pearls, their hair unbound, fanning out around the face in almost sensual abandon (Figs. 75, 76).[92] They figure the erotic plots of romance, types of the seductress and adulteress that people late eighteenth- and

Fig. 74. Anatomical model (18th cent.)

Fig. 75. (*Above*)
Anatomical model
(18th cent.)

Fig. 76. Anatomical model
(18th cent.)

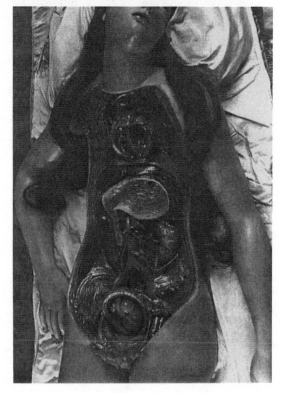

nineteenth-century novels, their sexual allure and power man-
aged by making them objects to be exposed and manipulated.
Their bodies are crafted to open, layer after layer, organ after organ, until a fetus is revealed (Fig. 77). They are always pregnant,
the mark of their sexual specificity. Such models were not only
used for anatomical study, but, like the one in Figure 78, were
commissioned in miniature for private collections as well. All of
these overdetermined, gendered, and heterosexual narratives for
coding the female body of anatomy—Eve, Venus, Natura, *la
grande horizontale*—ensure the terms on which the observing subject confronts sexual difference. To sum up, the woman's body is
sacrificed to fetal subjectivity in multiple ways: in absence; in the
visual schemas of the midwifery manuals; and, dissected and
opened, made manipulable, in the gendered plots of anatomical
illustration and sculpture.

It is often said that the laying open of the female body and its
dissection or division into parts is a strategy of mastery, and
Vesalius's frontispiece to the *Fabrica*, with its female corpse laid
out and opened to the beholder's eye, surrounded by a crowd
of men, is said to exemplify that strategy (Fig. 79).[93] Yet the division of the body into parts is not, of course, confined to the femi-
nine. Such fragmentation perhaps first appears in the wood,
terra-cotta, metal, and wax ex-voto figurines of body parts found
in Roman excavations, medieval churches, and pilgrimage sites
(Fig. 80) all over Europe and subsequently Latin America. The
great European eighteenth-century wax anatomical collections
are similarly made up of isolated body parts, not primarily of full
figures.[94] In museums in Bologna, Florence, and Vienna, huge
glass cases are filled with wax organs, muscles, and morcellated
body parts, mounted on hardwood plaques or turned pedestals
(Figs. 81–84), often framed by sculpted drapery that hints at a
not-so-veiled sadomasochism (Fig. 85).[95]

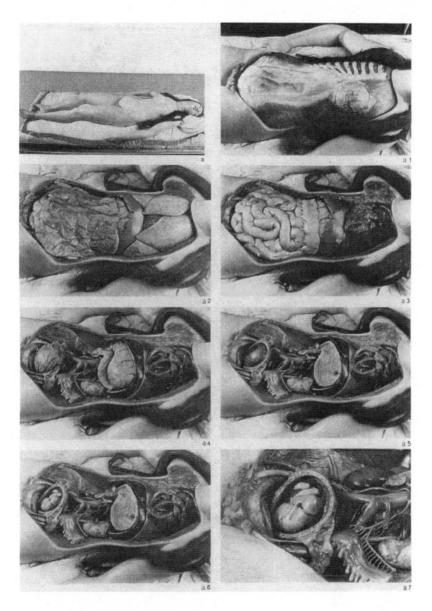

Fig. 77. Anatomical model (18th cent.)

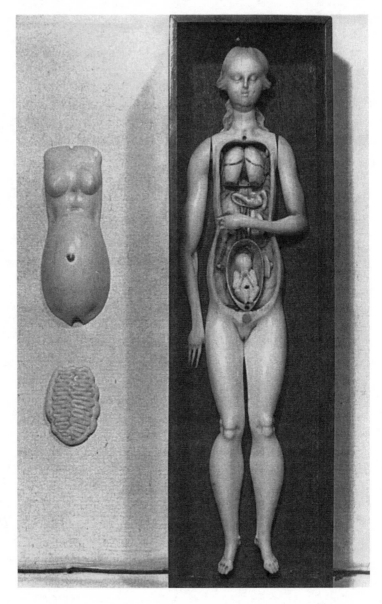

Fig. 78. Miniature anatomical model (18th cent.)

Fig. 79. Frontispiece of Vesalius, *De humani corporis fabrica* (1543)

Fig. 80. Etruscan Ex-
voto, uterus (3d cent.)

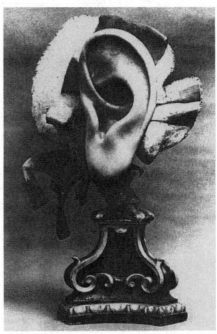

Figs. 81 & 82. Anatomical details (18th cent.)

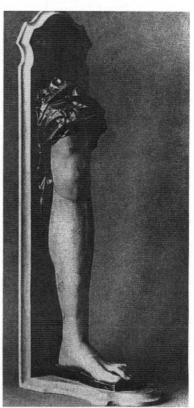

Figs. 83 & 84 (*Left and below right*)
Anatomical details (18th cent.)

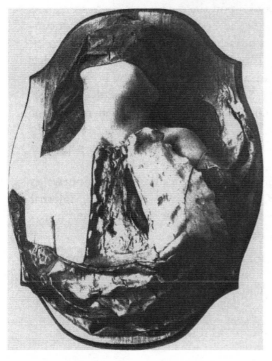

Fig. 85 (*Left*) Anatomical detail
(18th cent.)

In his discussion of the ex-voto tradition, David Freedberg argues that such sculpture is designed to "produce again, not reflect, illustrate, portray, or image. The striving for resemblance marks our attempts to make the absent present and the dead alive."[96] Freedberg's philosophical claims about the ex-voto are likewise suggestive for understanding anatomical sculpture. These individuated body parts, however, also bear witness to late seventeenth- and eighteenth-century mechanistic protocols of science, which were characteristic not only of the morcellated female body of anatomy but also of what I call the cultural logic of individualism. Knowledge was produced by the observation, dissection, and isolation of individual structures and by their taxonomic organization, a reductionist program exemplified by this quotation from the gentleman-scientist Robert Boyle: "I think the physician . . . is to look upon the patient's body as an engine that is out of order, but yet so constituted that, by his concurrence with the endeavors or rather the parts of the automaton itself, it may be brought to a better state."[97] Boyle's knowledge and figuration of the human body are fashioned by both anatomical practice and Royal Society scientific protocols based on demonstration, the mechanical device, and *a priori* notions of "parts." The physician is a mechanic or engineer, the body an engine or robot of assembled, isolatable moving parts and systems.

Knowledge of the body, then, in Foucault's arresting phrase, is "given only to that derisory, reductive, and already infernal knowledge that only wishes it dead."[98] Witness the recent introduction of a "Visible Man"

assembled digitally from thousands of X-ray, magnetic and photo images of cross sections of the body of Joseph Paul Jernigan, executed for killing a 75-year-old man during a burglary and who left his body to science. . . .

. . . The digitalized cadaver will be available free to anyone who ob-
tains permission from the library [National Library of Medicine] and
has the 15 gigabytes of storage space needed. . . .

. . . The library is spending $1.4 million on this project and [on] a
"Visible Woman." [99]

Twentieth-century biology, evolutionary theory, and medicine
remain heir to seventeenth-century reductionism and atomism
(Figs. 86–88) and continue to explain the "properties of collec-

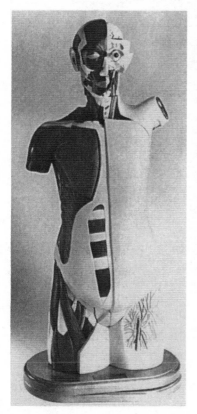

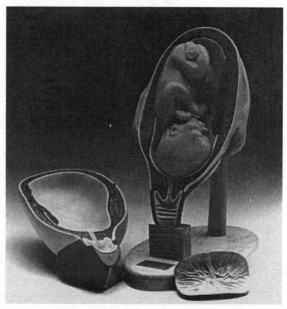

Fig. 86. (*Left*) Sexless multi-torso, anatomical
catalog (1994)

Fig. 87. (*Above*) Pregnancy insert, anatomical
catalog (1990)

Fig. 88.
Pregnant uterus
set, anatomical
catalog (1994)

tiv[ities] by the properties of assembled individuals."[100] Implied
in such a view is a notion of society as "a collection of causally
sovereign human individuals."[101]

In an interesting essay on the seventeenth-century anatomist
William Harvey, Luke Wilson describes as the epistemological
problem of anatomy that we see inside the body "only by vio-

lence and at the risk of pain. . . . Any glimpse of the interior of
the body is felt to invalidate it: a body whose interior is exposed
to the eye is always felt to be impaired or damaged."[102] Obstetrical visualizations challenge Wilson's powerful theorization of
anatomy, since such images inscribe both that body which is
opened to vision and at the same time another body, the fetal body, whole, impermeable, inviolate. The conventions for

inscribing obstetrical knowledge, then, allow for a double identificatory pleasure: identification with the immaculate, impene-
trable human individual, and the power/knowledge that comes
of knowing the body as an object of study. And that double pleasure is a profoundly gendered pleasure, as Figure 89, a most striking anatomical illustration of fetus and pregnant female body,

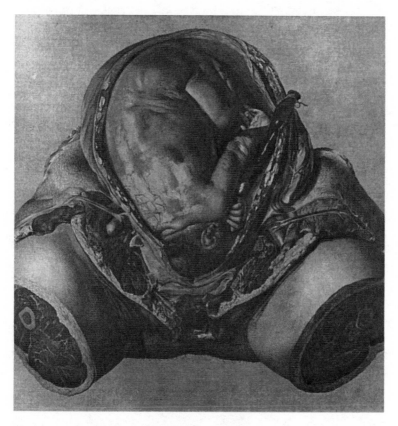

Fig. 89. Jan van Rymsdyk (artist), from William Hunter, *The Anatomy
of the Human Gravid Uterus* (1774)

demonstrates. Taken from William Hunter's obstetrical atlas *The Anatomy of the Human Gravid Uterus* (1774), it threatens the identificatory process I have analyzed.[103] Here is the conventional schema of anatomized uterus and fetus whole and complete, but the excess of anatomical detail in the sectioned thighs and the gratuitous mutilation of the genitals are incongruously juxtaposed with the glistening, impervious surface of the fetal body.[104] Hunter's engraving displays a kind of *folie de détail*, not only in the commonplace sense of "detail," but also in an etymological sense: the French *détailler* means "to cut in pieces."[105] His is a madness for the particular, a naturalism beyond denotative instrumentality in which the lavish format (an elephant folio to ac-

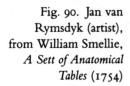

Fig. 90. Jan van Rymsdyk (artist), from William Smellie, *A Sett of Anatomical Tables* (1754)

commodate engravings to "natural size") and excessive detail collapse the conventional schema of obstetrical representation.

Hunter's image is unlike other contemporaneous renderings of obstetrical anatomy, such as Figure 90, from the well-known English man-midwife William Smellie's *A Sett of Anatomical Tables* (1754), which is more detailed than its predecessors but still schematically reminiscent of earlier fetal images; or Figure 91,

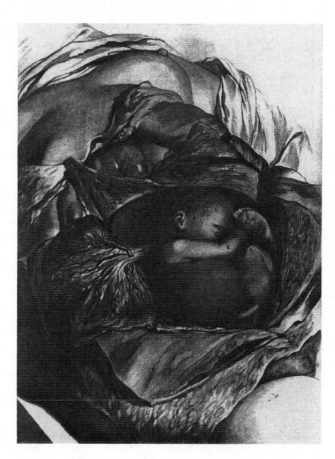

Fig. 91.
From William
Cowper, *The
Anatomy of
Human Bodies*
(1698)

from William Cowper's *The Anatomy of Human Bodies* (1698), and Figure 92, from Charles Jenty's obstetrical atlas *Demonstratio uteri praegnantis mulieris* (1759), with their more characteristic mitigation and eroticizing of anatomical violence by surrounding the opened body with drapery and leaving intact—recumbent in

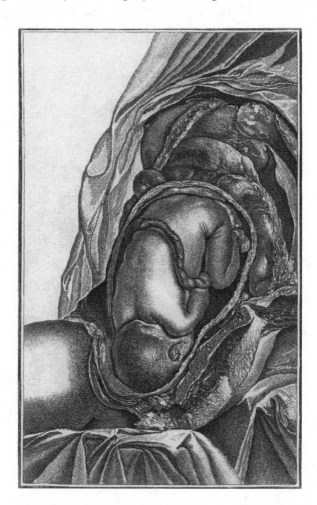

Fig. 92. From Charles N. Jenty, *Demonstratio uteri praegnantis mulieris* (1759)

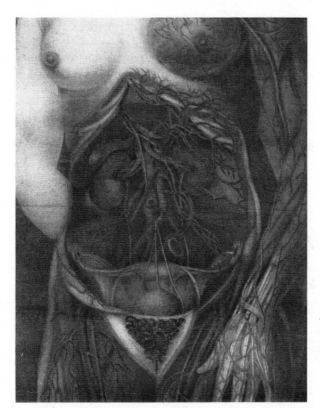

Fig. 93. From
Jacques Gautier
Dagoty, *Myologie
complette en
couleur et grandeur
naturelle* (1746)

Cowper, in profile in Jenty—the breast and pubic hair; or,
finally, Jacques Gautier Dagoty's remarkable colored mezzotint
(Fig. 93) from his *Myologie complette en couleur et grandeur naturelle*
(1746). Hunter's fetus is still an inviolate intact body, but it is
wedged unbearably not into a woman's body—Eve, Venus, odalisque—but, spectacularly, into meat.[106] The exorbitant detail that
renders anatomy butchery collapses the core schema by shifting
visual priorities: the viewer can no longer identify effortlessly
with the fetal body and encounter the female body as peripheral,
secondary, because that exorbitance refuses to allow the female

trunk to be mere setting. Instead the flesh asserts itself and threatens any logic of individualism by producing in the viewer an
uncomfortable knowledge not merely of his own carnality but
of the fragility of the separation of inside and outside, self and
other.[107]

∽

Although Hunter's *femina gravida* troubles conventional inscriptions of obstetrical knowledge, it is anomalous. Visualizations of knowledge about reproduction even in contemporary
obstetrical textbooks and models generally adhere to the hegemonic conventions I have analyzed. The most widely used stan-
dard obstetrical textbook in medical schools in the United States
and Canada, and one widely used abroad, is *Williams Obstetrics*,
first published in 1903 and now in its nineteenth edition; it begins by asserting: "The transcendent objective of obstetrics is that
every pregnancy be wanted, and that it culminate in a healthy
mother and a healthy baby."[108] In *Williams*, late twentieth-
century medical students learn that the "natural history of reproduction in our species has been obscured by social overlay, . . .
that genetically speaking we are still primitive hunter-gatherers"
(12), and that "women are physiologically ill-adapted to spend
the better part of their reproductive lives in the nonpregnant
state" (13−14).[109]

In a striking phrase reminiscent of early theories of fetal autonomy, aspiring obstetricians learn that "the fetus is the dynamic
force in the orchestration of its own destiny" (18); an early
twentieth-century medical man is quoted on menstruation: "pe-
riodic uterine hemorrhage is, in fact, one of the sacrifices which
women must offer at the altar of evolution and civilization" (98).
Apropos anatomical images, even as recently as the eighteenth
edition (1989), the book began by observing of Vesalius's *Fabrica*
that "the presentation of the gross human female reproductive

tract, in this masterpiece, was as we know it today" (23). In early editions of *Williams*, then entitled *Obstetrics: A Textbook for the Use of Students and Practitioners*, the frontispiece showed a pregnant woman's torso, with the caption reading: "Vertical Mesial Section Through Body of Woman Dying in Labour, with Unruptured Membranes Protruding from Vulva" (Fig. 94). But the sectioned fetus was apparently found offensive, and in the seventh edition (1936) an editor altered the illustration to depict the fetus intact. Finally the frontispiece disappeared altogether, relegated to the text to portray the complications of breech birth, but as a diagram rather than "a photograph-like drawing."[110]

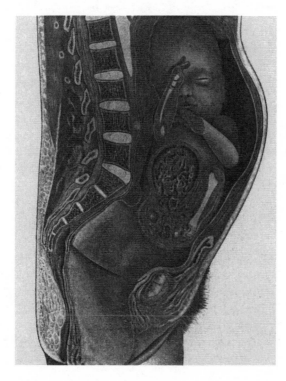

Fig. 94. From J. Whitridge Williams, *Obstetrics: A Textbook for the Use of Students and Practitioners* (1912)

In the most recent edition, the entire chapter on prenatal care or well-being, referred to as "prenatal surveillance," offers no illustration of an entire pregnant body. The draped crotches that are depicted, moreover, are primarily those of African American women, whose black bodies being delivered by white hands are inordinately visualized. Though made public in medical textbooks intended to educate doctors, ironically they themselves are too infrequently the recipients of what is taught through those books, from abortion and family planning services to prenatal care including high-tech medical techniques such as ultrasound. Their bodies, available in the large public hospitals that provide medical care to the majority of African Americans and other persons of color in the United States, are appropriated on behalf of medical pedagogy (Figs. 95, 96).

The first chapter of *Williams Obstetrics* closes with a remarkable paragraph that demonstrates both the continuing authority of rights claims and the influence of anti-abortion rhetoric:

The concept of the right of every child to be physically, mentally, and emotionally "well-born" is fundamental to human dignity. If obstetrics is to serve a role in the realization of this goal, the specialty must maintain and even extend its role in the control of population growth. The right to be "well-born" in its broadest sense is simply incompatible with unrestricted fertility. Yet our knowledge of the forces operative in the fluctuation and control of population is still rudimentary. This concept of obstetrics as a social as well as biological science impels us to accept a responsibility unprecedented in American medicine.[111]

Overtly, this paragraph takes no sides in the abortion debates; its ringing call for "wellness" (part of the jargon of modern preventive medicine), for "holistic" health—"physically, mentally, and emotionally"—for rights and human dignity, and for the social dimension of obstetrical practice on the face of it seems merely to represent a credo for the aspiring obstetrician. But the rhetorical

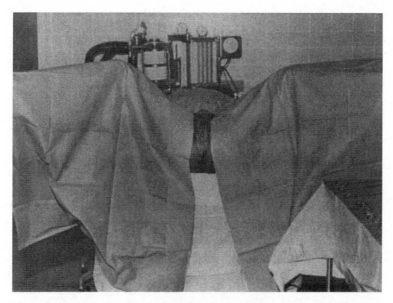

Fig. 95. From F. Gary Cunningham et al., *Williams Obstetrics* (1993)

Fig. 96. From Cun-
ningham et al., *Williams
Obstetrics*

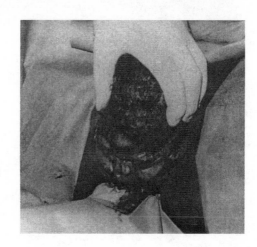

coupling of "children's rights" with birth, the double negative in the allusion to birth control—"incompatible," "unrestricted"— and the ambiguous assertion that the "forces operative in the fluctuation and control of population" are social rather than biological betray the political imaginary of the anti-abortion position. Population management is linked not to birth control or abortion, but to the personal and political—to rights, dignity, operative forces, responsibility.

Elsewhere, similarly, the fetus or neonate is a "child," an individualized subject. In illustrations of the various fetal presentations, whereas the pregnant woman's body is diagrammatically represented by a sketched skeletal pelvis, the fetus is represented quite differently, with cross-hatching to give shadow and texture to flesh, with humanizing details of hair, fingernails, and wrinkles, and including the marker of sexual difference, the penis (Figs. 97, 98). Important here are not the connotative details in and of

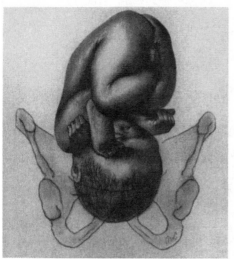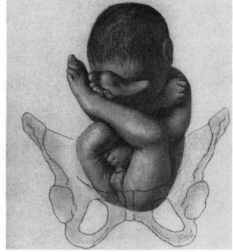

Figs. 97 & 98. From Cunningham et al., *Williams Obstetrics*

themselves, or the fact that the female body is rendered diagrammatically, but the *differential* modes of representing female and fetal bodies. Despite an increasing preference in medical illustration for the diagrammatic over the artistic in showing anatomic and pathological detail, the fetus remains subject to the canons of art. Two differing realist regimes, the diagrammatic and the naturalistic, present two quite different views of the "body's typicality," with quite different effects—effects at work from the very opening chapter of *Williams*.[112]

♂

The discourse of fetal personhood and the discourse of women's rights over their own bodies both presume a humanist, autonomous, observing subject. Contemporary fetal images produced by new visual apparatuses and technologies (sometimes termed "postmodern visualities"), however, are quite different from and noncontinuous with the modes of historical anatomical inscription I have been analyzing.[113] These simulated fetal images are subject to scatter, reflection, and other forms of artifactuality, including the "signature" of the traducer produced, for example, by ultrasonography.[114] Ultrasound, color Doppler sonography, and fetal laser imaging are not visual techniques that depend on an observing subject or its surrogate, the camera.[115] In other words, they are not analog media dependent on a point of view located in real space: they signify no "particularized individuality," they make no "deictic declaration: this is how I see."[116] Instead they are part of the burgeoning domain of alternative visualities—computer graphics, digital imaging and tomography, synthetic holography, robotic image recognition, ray tracing, texture mapping, magnetic resonance imaging (MRI), virtual reality simulators, and so forth: arrays of integers, noncontinuous imaging techniques that separate the visual from an observing subject and from all those older modes of seeing that rely on perspective, a fixed or mobile point of view. The visual apparatuses that produce fetal images—ultrasound, laser imaging—are part of what art historian Jonathan Crary has recently described as a "sweeping reconfiguration of relations between an observing subject and modes of representation that effectively nullifies most of the culturally established meaning of the terms *observer* and *representation*."[117]

Such image-producing technologies do not simply reproduce an unmediated truth that is "scientific knowledge"; the images generated by new visual apparatuses require reading and interpretation as fully as do older forms of representation.[118] The bewildering assortment of potential measurements; the interpretive problems and possibilities posed by digital image manipulation, color enhancement, and tonal gradation in relation to numerical differentials of data; the aesthetic abstraction of such images and their contextualization; the sheer diversity of visualization strategies and technologies—all witness the difficulty of interpreting such scientific inscriptions and their conventionalist character.[119] The problem of rendering these new visualizations readable is illustrated in the rhetoric and photographic juxtapositions needed to explain them to the public. Howard Sochurek's paean to the new visual apparatuses, *Medicine's New Vision*, for example, which first appeared as a *National Geographic* lead article, uses two strategies to "humanize" such highly abstract images.[120] The book unfolds as Sochurek's personal journey, a first-person account in which he describes undergoing successive tests and thereby demonstrating each procedure's "revolutionary," because noninvasive, character. He defines and illustrates the techniques with the "stories" of real people, their pictures juxtaposed with abstract and aestheticized visualizations. The photographs are mostly of children "saved" by the technique in question: a jaunty black toddler, whose tumor was found through ultrasound (Fig. 99), celebrates a birthday wearing a blue bow tie (Fig. 100);

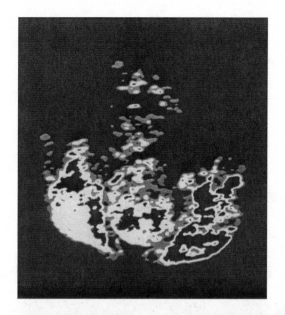

Fig. 99. (*Left*) Throat tumor (*National Geographic*, Jan. 1987)

Fig. 100. (*Below*) Joseph Ward (*National Geographic*, Jan. 1987)

a blond two-year-old saved by a PET (positron emission tomography) scan is shown with his physical therapist; a pregnant black woman, whose infertility was diagnosed using MRI, is pictured holding her pet duck; and there are testimonials and snapshots of men and women with captions declaring them "cured" of cancer. The color enhancement of images—"to help the reader share in and appreciate the way radiologists see"—is justified with examples: "the tumor, just a shadow, pops out in vivid red; the occlusion of the artery is easily seen; the yawning fetus comes alive." [121]

The expanding formalism of scientific visualization with its accompanying hermeneutic demands produces, in short, what might be termed a referential panic, a need for realist images in order to make "Joe's" and "Nathan's" and "Jimmy's" stories consumable. The new visualities unhinge the referent; the snapshots and captions seek to re-articulate them with Enlightenment and post-Enlightenment doctrines of rationality, which, in writing of Renaissance perspective, William Ivins termed the "rationalization of sight." [122] Sochurek's article and subsequent book, precisely by juxtaposing realist images with alternative forms of visual inscription, do not succeed in humanizing the new visualities, but rather underscore their *difference* from older modes of seeing. The color-enhanced yawning fetus (reproduced in black and white in Fig. 101) is described in the accompanying caption as a "first snapshot for the family album"; but far from coming alive, the image resembles a skull, a *memento mori*, the gaping grin of death.

New forms of visualization occupy at the very least an uneasy position in contemporary debates about reproduction. On the one hand, the New Right's deployment of these fetal images, as in *The Silent Scream*, the anti-abortion film purporting to record an abortion; the constitution of fetology and obstetrics around

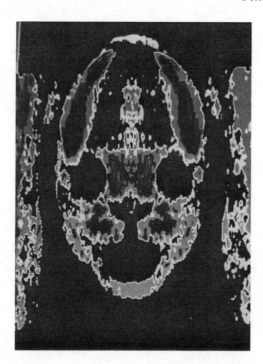

Fig. 101. Yawning
fetus (*National Geo-
graphic*, Jan. 1987)

the fetus as an individual in need of diagnosis and treatment;
medical, feminist, and cinematic discourses that fret over the
"bonding" mother; corporate appropriations of high-tech fetal
images in advertising aimed at upper-middle-class consumers
(Fig. 102); color-enhanced X-rays (Fig. 103) and computer images—all these factors collapse the mimetic and the simulated on
behalf of a humanist hermeneutics enabling to "pro-life" rhetoric
and the conservatism of the Rehnquist Supreme Court.[123] On
the other hand, the new visual apparatuses can potentially be har-
nessed to counter classical and Renaissance modes of representation, thus disrupting the cultural logic of individualism that relies

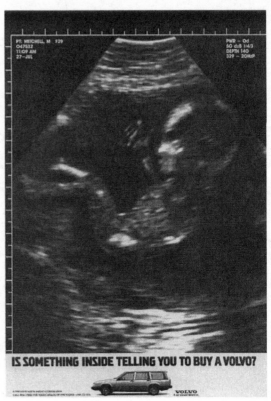

Fig. 102. Volvo advertisement, fetus at 12–16 weeks (1990)

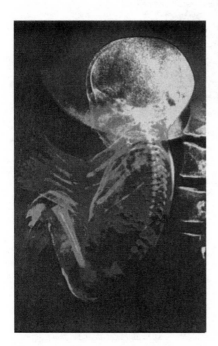

Fig. 103. Human fetus X-ray (ca. 1980)

on perspective—the rationalization of sight—and an observing subject interpellated to humanize such simulated images.

The right's insistent inscription of fetus as "baby" and feminist demands to restore the woman's body to obstetrical representations *both* display a profound humanist nostalgia for the realist image; at the same time, they perhaps seek to allay a profound anxiety about what constitutes "life" and the "individual" in the postmodern scientific environment of the cyborg. In the so-called life sciences, not only has the "ideal of the individual as an entity that may be treated as genetically uniform" been challenged, but "individuality" itself has been demonstrated to be "a derived character, approximated closely only in certain taxa." [124] Similarly, the power of reproductive technologies to disperse maternity materially as well as socially, through in vitro fertilization, artificial insemination, surrogacy, adoption, fosterage, and so forth, disrupts the singularity of maternal identity—a disruption enacted and enabled by cultural forms from coverage of the Baby M case to *Alien* and *Terminator 2* to the British cult film *Hardware*—and suggests the difficulty with identifying visual technologies merely as tools of institutional power. [125]

EPILOGUE

How we read the past—from the production of the rights-bearing subject to dominant modes of visualization that depend on a Cartesian observing subject—determines how we construct and interpret the present and affects profoundly the intelligibility of contemporary functionings of power. In the United States today, cultural studies are mounting an attack of the present on the rest of time;[1] this presentism, like the belligerent historicism of the academic right, threatens to obliterate not only the past but also the possibility of reading and acting politically in the present. Without historicizing hegemonic discursive and visual modes, we risk repeatedly reinscribing and thereby helplessly reproducing them. Early obstetrical illustration, Bologna's Museo ostetrico, and eighteenth-century anatomical sculpture and engraving are not merely antiquarian esoterica; rather, they constitute crucial political knowledge for the present. The adisciplinary environment of cultural studies has been preoccupied with periodization, the identification of ruptures and breaks, but grand claims for such shifts may be less significant than the analysis of modes of inscription: how inscriptions are assembled and manipulated, or, to return to Bruno Latour's formulation, how "groups of people argue with one another using paper, signs, prints and diagrams" so that "the weakest, by manipulating inscriptions of
` sorts obsessively and exclusively, become the strongest."[2]

This essay demonstrates how rights claims and individualism are entailed in fetal images, and suggests the consequences of that entailment for the abortion debates. Some readers might ask for more, for a "progressive" or "positive" alternative to the discourse of rights or the calls for restoring the maternal body to the fetal image. I have deliberately resisted such demands, remembering Adorno's observation that "repressive intolerance toward a thought not immediately accompanied by instructions for action is founded in fear."³ Instead of "solutions," which will inevitably differ in different communities, regions, temporal moments, and institutional circumstances, I offer what might be termed a politics of negation, since, again to use Adorno's words, "the seriousness of unswerving negation lies in its refusal to lend itself to sanctioning things as they are."⁴ A correlate of the argument I make here is, therefore, the vindication of intellectual labor and, by implication, the dismantling of the tired old division between theory and practice in feminist politics. The work of cultural critique, in my view, is not the presentation of alternative "progressive" narratives, but the adumbration of dystopia, of Adornian negative dialectics. Utopianism, as generations of readers of Sir Thomas More's *Utopia* have observed, slips all too easily into the totalitarian. Unlike Adorno, however, I do not believe that critical thinking is like bottles thrown into the sea by a particular thinker, presumably to wash up in some other place, in some other time, to be read and understood by some*one*. Instead, for me critical thinking is like plate tectonics: the earth moves ceaselessly, if almost always seemingly imperceptibly.

Reference Matter

NOTES

Prologue

1. Jacqueline Rose, *Sexuality in the Field of Vision* (London: Verso, 1986; rpt. 1991), 226.

2. *New York Times*, January 1, 1995, A1.

3. See particularly E. H. Gombrich, *Art and Illusion: A Study in the Psychology of Pictorial Representation* (Princeton, N.J.: Princeton University Press, 1960).

4. For an important discussion of the relation of early motion studies and cinema to the "modernist dynamic model of 'life' generated in physiology and medical science," see Lisa Cartwright, *Screening the Body: Tracing Medicine's Visual Culture* (Minneapolis: University of Minnesota Press, 1995).

5. On the "birth" of the homosexual, see Michel Foucault, *The History of Sexuality* (New York: Pantheon Books, 1978), vol. 1; K. Plummer, ed., *The Making of the Modern Homosexual* (Totowa, N.J.: Barnes & Noble, 1981); and David Halperin, *One Hundred Years of Homosexuality* (London: Routledge, 1990). For recent work on queer sexuality in the early modern period, see Alan Bray, *Homosexuality in Renaissance England* (London: Gay Men's Press, 1983); Bruce R. Smith, *Homosexual Desire in Shakespeare's England* (Chicago: University of Chicago Press, 1991); Jonathan Goldberg, *Sodometries* (Stanford, Calif.: Stanford University Press, 1992); and Jonathan Goldberg, ed., *Queering the Renaissance* (Durham, N.C.: Duke University Press, 1994).

6. On the origin of modern heterosexuality in the eighteenth cen-

tury, see Henry Abelove, "Some Speculations on the History of Sexual Intercourse During the Long Eighteenth Century in England," *Genders* 6 (1989): 125–30.

7. Judith Butler, *Bodies That Matter: On the Discursive Limits of Sex* (New York: Routledge, 1993), 116.

8. Thomas Laqueur, *Making Sex: Body and Gender from the Greeks to Freud* (Cambridge, Mass.: Harvard University Press, 1990), 14. Since it is not germane to my argument, I do not intend to take up Laqueur's claims for one sex in the early modern period. In brief, I agree with the critique advanced by Katharine Park and Robert Nye in their review of Laqueur in the *New Republic*, February 18, 1991, 53–57, though I would argue more vigorously for a blindness to figurative language in accounting for his provocative, but wrongheaded, argument.

9. Ian Hacking argues for what he calls "dynamic nominalism" in his tonic essay "Making Up People," in *Reconstructing Individualism: Autonomy, Individuality, and the Self in Western Thought*, ed. Thomas C. Heller, Morton Sosna, and David Wellbery (Stanford, Calif.: Stanford University Press, 1986), 222–36. The poststructuralist critique of both reference and the mirror model of literary production has been widely rehearsed. On this problem within the social study of science, see Bruno Latour's work, particularly *Science in Action* (Cambridge, Mass.: Harvard University Press, 1987); Donna Haraway, "Situated Knowledges: The Question in Feminism as a Site of Discourse on the Privilege of Partial Perspective," *Feminist Studies* 14 (1988): 575–600; Helen Longino, *Science as Social Knowledge: Values and Objectivity in Scientific Inquiry* (Princeton, N.J.: Princeton University Press, 1990); and N. Katherine Hayles, "Constrained Constructivism: Locating Scientific Inquiry in the Theatre of Representation," in *Realism and Representation*, ed. George Levine (Madison: University of Wisconsin Press, 1993), 27–33.

10. Rose, *Sexuality in the Field of Vision*, 231.

11. In the early modern period, before the constitution of a homosexual identity, when acts rather than bodies were the object of scorn, there is not yet a queer body invested with cultural meaning. See Bray, *Homosexuality in Renaissance England*. In the late twentieth century, the queer body, particularly if HIV-infected, is, of course, intensely overde-

termined, both differently from and in significantly similar ways to the female reproductive body.

12. For an early analysis of linguistic choices in the abortion debates, see Garrett Hardin, *Mandatory Motherhood: The True Meaning of "Right to Life"* (Boston: Beacon Press, 1974); and, more recently, Celeste Michelle Condit, *Decoding Abortion Rhetoric* (Chicago: University of Illinois Press, 1990). But see especially Barbara Johnson's analysis of rhetoric and abortion in "Apostrophe, Animation, and Abortion," *Diacritics* 16 (1986): 29−47.

13. On fetal imagery, see especially Zoë Sofia, "Exterminating Fetuses: Abortion, Disarmament, and the Sexo-Semiotics of Extraterrestrialism," *Diacritics* 14 (1984): 47−59; Rosalind Pollack Petchesky, "Foetal Images: The Power of Visual Culture in the Politics of Reproduction," in *Reproductive Technologies: Gender, Motherhood, and Medicine,* ed. Michelle Stanworth (Minneapolis: University of Minnesota Press, 1987), 57−80 (also included in part in her *Abortion and Woman's Choice: The State, Sexuality, and Reproductive Freedom,* rev. ed. [Boston: Northeastern University Press, 1990; 1st ed. 1984]); Carol A. Stabile, "Shooting the Mother: Fetal Photography and the Politics of Disappearance," *Camera Obscura* 28 (1992): 178−205; and, most recently, Lauren Berlant, "America, 'Fat,' the Fetus," *Boundary 2* 21 (1994): 145−95.

Fetal Positions

1. Bruno Latour, "Visualization and Cognition: Thinking with Eyes and Hands," in *Knowledge and Society: Studies in the Sociology of Culture Past and Present,* ed. Henrika Kuklick and Elizabeth Long (Greenwich, Conn.: JAI Press, 1986), 1−2. Subsequent in-text page references are to this essay.

2. On visualization in science, see Martin J. S. Rudwick's pioneering study "The Emergence of a Visual Language for Geological Science, 1760−1840," *History of Science* 14 (1976): 149−95; and Gordon Fyfe and John Law, eds., *Picturing Power: Visual Depiction and Social Relations, Sociological Review* Monograph 35 (London: Routledge, 1988).

3. For just such a grand conclusion, see, for example, Laqueur's "Sometime in the eighteenth century, sex as we know it was invented" (*Making Sex,* 149).

4. See among others Petchesky, "Foetal Images"; Stabile, "Shooting the Mother"; and Barbara Duden, *Disembodied Women* (Cambridge, Mass.: Harvard University Press, 1993).

5. April 30, 1965; the series subsequently appeared, with numerous additional images, in Nilsson's *A Child Is Born* (1965; New York: Doubleday, 1990).

6. Figure 4 is described on the front cover as a "living 18-week-old fetus," but inside (3) it is explained that the "embryo was photographed just after it had to be surgically removed . . . [and] did not survive."

7. On Nilsson's equivocations between "living fetus" and specimens, see Alice E. Adams, *Reproducing the Womb: Images of Childbirth in Science, Feminist Theory, and Literature* (Ithaca, N.Y.: Cornell University Press, 1994), 141–42.

8. See Michael Lynch, "Discipline and the Material Form of Images: An Analysis of Scientific Visibility," *Social Studies of Science* 15 (1985): 37.

9. *The Camera* (New York: Time-Life Books, 1970), 52.

10. See Michel Foucault, *The Birth of the Clinic* (New York: Pantheon Books, 1973); and, more recently, Sarah Kember, "Medical Diagnostic Imaging: The Geometry of Chaos," *New Formations* 15 (1991): 55–66; and Janelle S. Taylor, "The Public Fetus and the Family Car: From Abortion Politics to a Volvo Advertisement," *Public Culture* 4 (1992): 67–70.

11. Lynch, "Discipline and the Material Form of Images," 44.

12. See Condit, *Decoding Abortion Rhetoric*, who points out that antiabortion slide shows, when using fetal images of early development, label them prominently "baby," surround them with images of more developed fetuses and of babies, and accompany them with text that attributes babylike features to them such as thumb-sucking, brain activity, and so on (83). Often they begin with footage or images of babies "to establish the continuity of the life forms" (85).

13. On the microphysical body and scientific omniscience, see chapter 4 of Lisa Cartwright's discussion of microscopy in *Screening the Body*.

14. See Petchesky, *Abortion and Woman's Choice*, in which she points out the further irony that "it is a propagandistic tour de force to have

taken the notion of 'personhood' (a metaphorical, moral idea) and translated it into a series of arresting visual images that are utterly physiological and often just plain morbid . . . fetal 'personhood' in the guise of human embodiment (in contrast, note, to 'ensoulment')" (338–39). See also Sarah Franklin, "Fetal Fascinations: New Dimensions to the Medical-Scientific Construction of Fetal Personhood," in *Off-Centre: Feminism and Cultural Studies*, ed. Sarah Franklin, Celia Lury, and Jackie Stacey (New York: HarperCollins, 1991), 190–205.

15. On spectacle as a specious form of the sacred, see Guy Debord, *Comments on the Society of the Spectacle* (New York: Verso, 1990).

16. Lennart Nilsson and Jan Lindberg, *Behold Man: A Photographic Journey of Discovery Inside the Body* (Boston: Little, Brown, 1974), 54.

17. Jon Darius, *Beyond Vision* (Oxford: Oxford University Press, 1984), 120.

18. On the shift from the so-called medical model to the rights model and the politics of abortion, see Faye Ginsburg, *Contested Lives: The Abortion Debate in an American Community* (Berkeley and Los Angeles: University of California Press, 1989), 39ff.

19. In addition to fetal rights and women's rights, the family has entered the ring in the competition for rights. See *The Terrible Choice: The Abortion Dilemma*, ed. and written by the Joseph P. Kennedy Jr. Foundation, with the help of Robert E. Cooke et al. (New York: Bantam Books, 1968), 73, quoting Dr. Sophia J. Kleegman in support of abortion on "the rights of the family to be allowed to maintain some stability and to offer a healthy emotional climate for the child."

20. Kristin Luker, *Abortion and the Politics of Motherhood* (Berkeley and Los Angeles: University of California Press, 1984), 5 (italics hers). On the issue of rights and abortion in Britain, see the series of articles on the Alton Bill, feminism, and science in Franklin, Lury, and Stacey, eds., *Off-Centre*.

21. Quoted in Petchesky, *Abortion and Woman's Choice*, 335. See also J. Douglas Butler and David F. Walbert, eds., *Abortion, Medicine, and the Law*, 3d ed. (New York: Facts on File, 1986).

22. Quoted in Susan Faludi, *Backlash: The Undeclared War Against American Women* (New York: Crown Books, 1991), 406. Faludi provides a useful survey of this phenomenon; but see also Petchesky, *Abortion and Woman's Choice*.

23. Reported in *New York Magazine*, April 24, 1989, 48−51, in an article by Peg Tyre.

24. Quoted in Anthony Lewis, "Right to Life," *New York Times*, March 12, 1993, A29.

25. Quoted in the *New Paper* (published by Phoenix Communications Group, Boston), January 1991.

26. Quoted in Luker, *Abortion and the Politics of Motherhood*, 150.

27. Quoted in Ginsburg, *Contested Lives*, 106.

28. See ibid., 106ff., on the intercutting in anti-abortion films, slide shows, and videos of footage of Southeast Asian children during the Vietnam war accompanied by claims that saline solution must "feel like napalm."

29. For a discussion of the relationship between individualism and the nation, see, for example, Louis Dumont, *Essays on Individualism* (Chicago: University of Chicago Press, 1986), 10.

30. *Christianity Today*, July 12, 1985, 40.

31. Ibid. See the article by Carol Vinzant in the May 1993 issue of *Spy* on the alleged theft and use of aborted fetuses by the Right-to-Life movement.

32. Mary Poovey, "The Abortion Question and the Death of Man," in *Feminists Theorize the Political*, ed. Judith Butler and Joan Wallach Scott (New York: Routledge, 1992), 241. For other recent discussions of "rights," see Martha Minow, "Interpreting Rights: An Essay for Robert Cover," *Yale Law Journal* 96 (1987): 1860−1915; Patricia Williams, *The Alchemy of Race and Rights* (Cambridge, Mass.: Harvard University Press, 1991); and, in Britain, Deborah Lynn Steinberg, "Adversarial Politics: The Legal Construction of Abortion," in Franklin, Lury, and Stacey, eds., *Off-Centre*, 175−89.

33. As a working definition of "individualism," I use *The Blackwell Companion to the Enlightenment* (Oxford: Blackwell, 1991): the "doctrine that privileges the individual human person, his rights and needs, over and above all social institutions, and maintains that the individual can be considered independently of any social grouping or framework" (242).

34. For a consideration of visual culture in the Enlightenment, see Barbara Maria Stafford's ambitious and lavishly illustrated *Body Criticism: Imaging the Unseen in Enlightenment Art and Medicine* (Cambridge, Mass.:

MIT Press, 1991). Zoë Sofia uses the word "personhood" in her "Exterminating Fetuses."

35. See Stabile, "Shooting the Mother," 179−80.

36. Petchesky, "Foetal Images," 61.

37. Sofia, "Exterminating Fetuses," 49.

38. Duden, *Disembodied Women*, 67; see also Berlant, "America, 'Fat,' the Fetus," on the "convergence of mass culture and mass nationality [in] an image of citizenship as a kind of iconic superpersonhood, of which the fetus is the most perfect unbroken example," 148.

39. See John Tagg's discussion of Roland Barthes's *Camera Lucida* and the indexical character of the photograph in *The Burden of Representation* (Minneapolis: University of Minnesota Press, 1993), 1ff. On the important distinction between older static investigatory modes of imaging the body, such as anatomy and photography, and the motion study and cinema, see Cartwright, *Screening the Body*.

40. See Jeremy M. Norman, ed., *Morton's Medical Bibliography: An Annotated Check-List of Texts Illustrating the History of Medicine*, 5th ed. (Brookfield, Vt.: Scholar Press, 1991; first published by L. T. Morton, 1943). For an invaluable guide to the illustration of obstetrics and gynecology generally, see Harold Speert, *Iconographia Gyniatrica: A Pictorial History of Gynecology and Obstetrics* (Philadelphia: F. A. Davis, 1973).

41. The full text of Soranus's *Gynaecia* was published in Greek by F. R. Dietz, *De arte obstetricia morbisque mulierum quae supersunt* (Regimontii: Graefe & Unzer, 1838), from manuscripts then in Paris and Rome. Manuscripts containing Muscio's treatise based on a short version of Soranus that is apparently no longer extant can be found in several collections in Europe, including the Bibliothèque Nationale in Paris, the Bibliothèque Royale Albert Iᵉʳ in Brussels, the Vatican, and at least one German collection. Though the prominent genitals of the Muscio figures reproduced in Fig. 16 make them appear to be sexed male, the erect, disproportionate penises are schoolboy doodling added in the twentieth century. In a recent catalog published by the Bibliothèque Nationale on the history of medicine in which these illustrations appear, no mention is made of these alterations. Perhaps the representation of the male fetus in obstetrical illustration is so conventional as to seem not to require commentary. Though early obstetrical illustrations

based on Soranus and the Muscio manuscripts suggest a variety of fetal positions, actually some 96 percent of fetal presentations are vertex (head down).

42. The figures were also reproduced in Italian manuals such as Scipione Mercurio's *La comare o riccoglitrice* (Venice, 1596).

43. In his *History and Bibliography of Anatomic Illustration*, trans. and ed. Mortimer Frank (Chicago: University of Chicago Press, 1920; 1st ed. Leipzig, 1852), Ludwig Choulant maintains that the Soranian illustrations came to Rösslin via the Heidelberg Codex, now at the Vatican. The Rösslin woodcuts are by Erhard Schön.

44. See Charles W. Bodemer, "Embryological Thought in Seventeenth-Century England," in *Medical Investigations in Seventeenth-Century England* (Los Angeles: Clark Memorial Library Publications, 1968); also Ludmilla Jordanova, "Gender, Generation, and Science: William Hunter's Obstetrical Atlas," in *William Hunter and the Eighteenth-Century Medical World*, ed. W. F. Bynum and Roy Porter (Cambridge: Cambridge University Press, 1985), 385–412.

45. Robert A. Erickson, "'The Books of Generation': Some Observations on the Style of British Midwife Books, 1671–1764," in *Sexuality in Eighteenth-Century Britain*, ed. Paul-Gabriel Bouce (Manchester: Manchester University Press, 1982), 77. See also Elizabeth B. Gasking, *Investigations into Generation, 1651–1828* (London: Hutchinson, 1967); Charles Bodemer's review of medieval and early modern theories of the uterus in "Historical Interpretations of the Human Uterus and *Cervix Uteri*," in *The Biology of the Cervix*, ed. Richard J. Blandau and Kamran Moghbissi (Chicago: University of Chicago Press, 1973), 1–11; and Audrey Eccles, *Obstetrics and Gynaecology in Tudor and Stuart England* (Kent, Ohio: Kent State University Press, 1982), 55–56.

46. On how pseudoscience is claimed as rational and on science itself as a major source of rejected knowledge, see W. M. Williams and R. J. Frankenberg, *On the Margins of Science: The Social Construction of Rejected Knowledge*, Sociological Review Monograph 27 (Keele, Eng.: University of Keele, 1979).

47. See, for example, Loris Premuda's discussion of this drawing in *Storia dell'iconografia anatomica* (Milan: Aldo Martello, 1957), 84.

48. Norman Bryson, *Vision and Painting: The Logic of the Gaze* (New

Haven, Conn.: Yale University Press, 1983), 153. Subsequent page references to this book are cited in the text.

49. I am grateful to Margreta de Grazia for insisting in an epistolary conversation that I be more explicit in distinguishing how the early and late images mean.

50. On wax anatomical sculpture, see *La ceroplastica nella scienza e nell'arte* (Florence: Olschki, 1977), proceedings of the conference of that name held in Florence in 1975; and C. J. S. Thompson, "Anatomical Manikins," *Journal of Anatomy* 59 (1925): 442–45.

51. On early modern collecting and the idea of the museum, see Paula Findlen, "The Museum: Its Classical Etymology and Renaissance Genealogy," *Journal of the History of Collections* 1 (1989): 59–78. On the logic of the museum, see Ludmilla Jordanova, "Museums: Representing the Real?" in Levine, ed., *Realism and Representation*, 255–78; on the gendering of collections and exhibits, see Gaby Porter, "Partial Truths," in *Museum Languages: Objects and Texts* (Leicester: Leicester University Press, 1991), 101–18.

52. Although it is generally said that surgery was a less prestigious practice in early modern Europe than other medical specialties, Galli was well respected. His career in fact helped to establish surgery and obstetrics in Bologna's medical school. On Galli's role in establishing this collection, see Marco Bortolotti and Viviana Lanzarini, eds., *Ars obstetricia bononiensis* (Bologna: CLUEB, 1988), 16 (Museo ostetrico catalog). On the scholarly overstatement of the conflict between physician and surgeon except in London and Paris, see C. D. O'Malley, "Medical Education During the Renaissance," *The History of Medical Education*, ed. C. D. O'Malley (Berkeley and Los Angeles: University of California Press, 1970), 100.

53. See Luigi Belloni, "Italian Medical Education After 1600," in O'Malley, ed., *History of Medical Education*, 105–19.

54. Interestingly, Galli is believed to have been prompted to commission his obstetrical models on the basis of illustrations in the well-known manuals of François Mauriceau, *Traité des maladies des femmes grosses* (Paris, 1740; 1st ed. 1668); and Hendrik van Deventer, *Operationes chirurgicae novum lumen exhibentes obstetricantibus* (Leiden, 1701). See Bortolotti and Lanzarini, eds., *Ars obstetricia*, 68.

55. The clay models are believed to have been sculpted by Giovanni Battista Sandri, the woodwork by Antonio Cartolari, and a few additional wax models by the Toselli brothers. On the crafting of the collection, see Marco Bortolotti's essay in Bortolotti and Lanzarini, eds., *Ars obstetricia*, esp. 21–22.

56. Quoted in Bartolotti and Lanzarini, eds., *Ars obstetricia*, 17.

57. On the man-midwife, see Adrian Wilson, "William Hunter and the Varieties of Man-Midwifery," in Bynum and Porter, eds., *William Hunter and the Eighteenth-Century Medical World*, 344–69.

58. See Eli F. Heckscher, *Mercantilism* (London: George Allen & Unwin, 1934), 2:158–61; and Othmar Keel, "The Politics of Health and the Institutionalization of Clinical Practices in Europe in the Second Half of the Eighteenth Century," in Bynum and Porter, eds., *William Hunter and the Eighteenth-Century Medical World*, 214ff. On populationism and public health, see also George Rosen, *From Medical Police to Social Medicine: Essays on the History of Health Care* (New York: Science History Publications, 1974), 120ff.

59. For a salutary challenge to prevailing views of women's management of women's health, with important implications for such claims throughout the early modern period, see Monica Green, "Women's Medical Practice and Health Care in Medieval Europe," *Signs* 14 (1989): 434–73. Although Green's evidence suggests that women certainly practiced medicine in a variety of capacities, and that men were involved in women's health care, she accepts that women seem to have been the primary attendants at normal childbirths. On women health practitioners, see also Margaret Pelling and Charles Webster, "Medical Practitioners," in *Health, Medicine, and Mortality in the Sixteenth Century*, ed. Charles Webster (Cambridge: Cambridge University Press, 1979), 165–235; and Green's review of the literature.

60. For Italy, see Claudia Pancino, "La comare levatrice: Crisi di un mestiere nel XVIII secolo," *Storia e società* 13 (1981): 593–638; and her essay "L'ostetricia del Settecento e la scuola bolognese di Antonio Galli," in Bortolotti and Lanzarini, eds., *Ars obstetricia*. See also M. G. Nardi, "La fondazione in Italia delle prime scuole ostetriche e la contesa priorità dell'istituzione dell'insegnamento ufficiale dell'ostetricia nelle università di Bologna e di Firenze," *Rivista italiana di ginecologia* 38

(1955): 177–84. For England, see Hilda Smith, "Gynecology and Ideology in Seventeenth-Century England," in *Liberating Women's History: Theoretical and Critical Essays*, ed. Berenice A. Carroll (Urbana: University of Illinois Press, 1976), 107–13; Thomas G. Benedek, "The Changing Relationship Between Midwives and Physicians During the Renaissance," *Bulletin of the History of Medicine* 51 (1977): 550–64; and Richard Wilson, "Observations on English Bodies: Licensing Maternity in Shakespeare's Late Plays," in *Will Power* (Detroit: Wayne State University Press, 1993), 158–83. For a survey of midwifery as a progression toward science and the professionalization of obstetrics, see I. S. Cutter and H. R. Viets, *A Short History of Midwifery* (Philadelphia: Saunders, 1964); and H. Spencer, *History of British Midwifery, 1650–1800* (London: Bale & Danielsson, 1927). For the United States, see Harold Speert, *Obstetrics and Gynecology in America: A History* (Chicago: American College of Obstetricians and Gynecologists, 1980). For the feminist challenge to the "great men" theory of midwifery, see Jean Donnison, *Midwives and Medical Men: A History of Inter-Professional Rivalries and Women's Rights* (New York: Schocken Books, 1977); and Ann Oakley, *Women Confined: Towards a Sociology of Childbirth* (New York: Schocken Books, 1980).

61. Barber-surgeons occupied a place somewhere in between and included women among their ranks. On medical practice as a site of contest among a variety of practitioners, male and female, see Green, "Women's Medical Practice," 446–52.

62. Biagi claims in her introduction to *Medicina per le donne nel Cinquecento*, ed. Maria Luisa Altieri Biagi, Clemente Mazzotta, Angela Chiantera, and Paola Altieri (Turin: UTET, 1992), 13, that although sixteenth-century manuals repeatedly claim "comare, raccoglitrici, mammane e levatrice" as their audience, they must have been read mainly by doctors and those bureaucrats seeking to control midwives on behalf of public health. Issues of literacy certainly complicate how we are to understand these manuals and their purposes in the early modern period, though urban midwives, to whom such manuals would have been most widely marketed, would have had a much higher rate of literacy than their rural counterparts. On questions of literacy and address in medical manuals concerned with women's reproductive health, see Green, "Women's Medical Practice," 453–69.

63. The rhetoric of blame and midwifery seems to have been a topos, though it may have expanded in the mid–eighteenth century. Subsequent demographic work demonstrates that this professionalization of obstetrics and the suppression of midwifery did not reduce the rate of infant mortality. See Thomas McKeown, *Laumento della popu-lazione nell'età moderna* (Milan: Feltrinelli, 1976; rpt. 1979), 145–48; cited by Pancino in her essay "L'ostetricia del Settecento." Pancino contends that the Museo ostetrico and the new schools were established out of a concern for infant mortality among city officials and doctors, whose motives, she claims, were to improve the techniques of mid-wives and to educate young women, not to repress popular medicine. Certainly this was an important function and aim of the collection, manuals, and newly instituted courses, but such expressed motives can-not be taken simply at face value. Ultimately these changes did result in the professionalization of obstetrics and its increasing domination by men. Furthermore, Pancino suggests that the use of pedagogic models such as are found in the Museo ostetrico represented a way of teach-ing illiterate women practitioners. But given the presence of similar anatomical models for the express use of literate medical men and the continuing debates over women's literacy, these obstetrical models sug-gest more complex motives and effects. The literature on the medical-ization of childbirth in the United States is enormous; see especially Richard Wertz and Dorothy C. Wertz, *Lying-In: A History of Childbirth in America* (New York: Free Press, 1977); James Mohr, *Abortion in Amer-ica: The Origins and Evolution of National Policy* (New York: Oxford Uni-versity Press, 1978); and Frances E. Kobrin, "The American Midwife Controversy: A Crisis of Professionalization," in *Women and Health in America*, ed. Judith Walzer Leavitt (Madison: University of Wisconsin Press, 1984), 318–26. On the British debate, see Wendy Fyfe, "Abor-tion Acts: 1803 to 1967," in Franklin, Lury, and Stacey, eds., *Off-Centre*, 160–74. For Eastern Europe, see Alena Heitlinger, *Reproduction, Medi-cine, and the Socialist State* (London: Macmillan, 1987); and Eleonora Zielnska, "Recent Trends in Abortion Legislation in Eastern Europe, with Particular Reference to Poland," *Criminal-Law-Forum* 4 (1993): 47–93. For an interesting Foucauldian account that argues obstetrics

has recently undergone a shift as profound as its professionalization and masculinization in the late eighteenth and nineteenth centuries, a shift based on the will to know, predict, control, and rationalize, see William Ray Arney, *Power and the Profession of Obstetrics* (Chicago: University of Chicago Press, 1982), and Robbie E. Davis-Floyd, "Obstetric Training as a Rite of Passage," *Medical Anthropology Quarterly* 1 (1987): 288–318.

64. For example, the so-called *capotomobolo*, a fantasized somersault the fetus was believed to make *in utero* immediately before birth.

65. The three "flexible" fetuses Galli used to illustrate correct birthing procedures in the transparent uterus have been lost, but they are listed in the inventory.

66. On the frequently recounted story of the forceps, see John Hobson Aveling, *The Chamberlens and the Midwifery Forceps: Memorials of the Family and an Essay on the Invention of the Instrument* (London: J. H. Churchill, 1882); Walter Radcliffe, *The Secret Instrument: The Birth of the Midwifery Forceps* (London: Heinemann, 1947); and Cutter and Viets, *Short History of Midwifery*, 44–69.

67. See Susan M. Pearce, "Collecting Reconsidered," in *Museum Languages*, 135–53, on serial relationships, collections, and the production of knowledge; see also Condit, *Decoding Abortion Rhetoric*, 85.

68. The video presents the Museo ostetrico via a narrator, Renzo Predi, a professor of statistics and demographic analysis at the University of Bologna. He stands not in the museum proper, but in what is called the anatomical theater, in all its ornate glory, next to the lecture rostrum with its carved wooden canopy supported by two *scorticati* or "flayed ones" (1734) by Ercole Lelli, an early anatomical sculptor with whom Giovanni Manzolini studied. The collection is thereby imbued with the prestige of the *teatro anatomico* and the Bolognese anatomical academic tradition, and the name "anatomical theater," still in use, emphasizes the spectacular dimension of early anatomical study and pedagogy. See Giovanna Ferrari, "Public Anatomy Lessons and the Carnival: The Anatomy Theatre of Bologna," *Past and Present*, no. 117 (1987): 50–106. In the video, the commissioning and assembling of the collection and the history of birthing are presented as a narrative of progress moving toward the modern, overcoming infant mortality, instituting public health, professionalizing obstetrics—this despite demographic

study which has shown that these "improvements" did not in fact change the infant mortality rate until well into the twentieth century. Finally, early obstetrics is presented via Scipione (sometimes cited as Girolamo) Mercurio's *La comare*, a manual whose title itself does cultural work. *Comare* was the common colloquial term for midwives (It. *levatrice*), but meant literally "gossips" and thus labels midwives as older, gabbing, meddling women. See the recent collection that includes Mercurio and another manual by Giovanni Marinello, *Medicina per le donne nel Cinquecento*, ed. Maria Luisa Altieri Biagi, Clemente Mazzotta, Angela Chiantera, and Paola Altieri (Turin: UTET, 1992).

69. For a powerful account of the privileging of vision since Plato, see Martin Jay, *Downcast Eyes: The Denigration of Vision in French Twentieth-Century Thought* (Berkeley and Los Angeles: University of California Press, 1993); for a review of the literature on perspective, see particularly 69–82. On the shift from perspective as a geometric practice to perspective as a "metaphor, a powerful concept for ordering our perception and accounting for our subjectivity" (xi), see James Elkins, *The Poetics of Perspective* (Ithaca, N.Y.: Cornell University Press, 1994).

70. Brian Rotman, *Signifying Nothing: The Semiotics of Zero* (Stanford, Calif.: Stanford University Press, 1987), 19. See also Peter Galassi, *Before Photography: Painting and the Invention of Photography* (New York: Museum of Modern Art, 1981).

71. Michael Baxandall, *The Limewood Sculptors of Renaissance Germany* (New Haven, Conn.: Yale University Press, 1980), 153.

72. See Hal Foster's collection *Vision and Visuality*, Dia Art Foundation Discussions in Contemporary Culture 2 (1988). On the Enlightenment and the shift to a selfhood that is psychological and personal, see Roy Porter, "The Enlightenment in England," in *The Enlightenment in National Context*, ed. Roy Porter and Mikulas Teich (Cambridge: Cambridge University Press, 1981), 14; and Michel Foucault, "The Subject and Power," *Critical Inquiry* 8 (1982): 777–95. For an early example of this rhetoric of individualism in the abortion debate, consider Pearl Buck's foreword to *The Terrible Choice: The Abortion Dilemma* (New York: Bantam Books, 1968), a collection based on a Harvard Divinity School conference: "Unless molested, unless life-support is interfered with or withdrawn, human fetal tissue has the potential or capability—indeed

the likelihood—of developing into an individual like ourselves, to be like us, to be in our own image, to be human" (2).

73. Donna Haraway, *Primate Visions* (New York: Routledge, 1989), 353. Petchesky, in *Abortion and Woman's Choice*, also notes in passing that "the free-floating fetus merely extends to gestation the Hobbesian view of human beings as disconnected, solitary individuals, paradoxically helpless and autonomous at the same time. It is this abstract individualism effacing the pregnant woman and the fetus's dependence on her, that gives the fetal image its symbolic transparency" (xiv). On the psychoanalytic dimension of pregnancy's doubled subjectivity, see Eugénie Lemoine-Luccioni, *Partège des femmes* (Paris: Editions du Seuil, 1976); Julia Kristeva, "Women's Time," trans. Alice Jardine and Harry Blake, in *Feminist Theory: A Critique of Ideology*, ed. Nannerl O. Keohane, Michelle B. Rosaldo, and Barbara C. Gelpi (Chicago: University of Chicago Press, 1982), 31–54; and idem, "Stabat Mater," in *The Female Body in Western Culture: Contemporary Perspectives*, ed. Susan Rubin Suleiman (Cambridge, Mass.: Harvard University Press, 1986), 99–118. See also Emily Martin, "Pregnancy, Labor, and Body Image in the United States," *Social Science and Medicine* 11 (1984): 1201–6; and Iris Marion Young, "Pregnant Embodiment: Subjectivity and Alienation," *Journal of Medicine and Philosophy* 9 (1984): 45–62.

74. Foster, *Vision and Visuality*, x.

75. Raymond Williams, *Keywords* (Glasgow: Fontana, 1976), 133. See also Peter Stallybrass, "Shakespeare, the Individual, and the Text," in *Cultural Studies*, ed. Lawrence Grossberg, Cary Nelson, and Paula Treichler (New York: Routledge, 1992), 593–612, for a keen discussion of this shift in meaning in the context of early modern English culture and politics.

76. C. B. Macpherson, *The Political Theory of Possessive Individualism* (Oxford: Oxford University Press, 1962; rpt. 1979). See also Sharon Marcus, "Placing *Rosemary's Baby*," *differences* 5 (1994): 121–53, in which she situates medical "surveillance" in the context of the post–World War II anxiety about privacy and argues that "women and their bodies were seen as outside the pale of possessive individualism" (140). For a survey of the development of notions of the self and individualism, and especially of the vexed relation of Enlightenment

thinkers to materialism, see Charles Taylor, *Sources of the Self* (Cambridge, Mass.: Harvard University Press, 1989).

77. John Locke, *The Second Treatise of Government*, ed. Thomas P. Peardon (New York: Liberal Arts Press, 1952), 16, 17, 41.

78. Haraway, *Primate Visions*, 353.

79. Jordanova, "Gender, Generation, and Science," 406–7.

80. Quoted in Paula Treichler, "Feminism, Medicine, and the Meaning of Childbirth," in *Body/Politics*, ed. Mary Jacobus, Evelyn Fox Keller, and Sally Shuttleworth (New York: Routledge, 1990), 135. Treichler also observes that birth is sometimes represented as the accomplishment of the physician. On this reemergence of the fetus as individual agent, see Sarah Franklin's discussion of "fetology" in "Fetal Fascinations." Franklin's claim that fetal agency is a "significant departure from previous medical and scientific constructions of fetal life" (194) is belied by early theories of preformationism (see page 66 above, and notes 44 and 45). On recent clinical work demonstrating that labor and uterine contraction are not simply involuntary, see Martin, "Pregnancy, Labor, and Body Image."

81. On fetology, see E. Peter Volpe, *Patient in the Womb* (Macon, Ga.: Mercer University Press, 1984); Asini Kurjak, *The Fetus as a Patient* (Amsterdam: Excerpta Medica, 1985); and especially Michael R. Harrison, Mitchell S. Golbus, and Roy A. Filly, eds., *The Unborn Patient: Prenatal Diagnosis and Treatment*, 2d ed. (Philadelphia: W. B. Saunders, 1990).

82. Michael Harrison, "Unborn: Historical Perspective of the Fetus as a Patient," *Pharos* 45 (1982): 22, 24; quoted in Robyn Rowland, *Living Laboratories: Women and Reproductive Technologies* (Bloomington: Indiana University Press, 1992), 120. Harrison ends his cursory historical survey of recent developments in fetology with a highly problematic and rhetorically revealing wish: "There is promise that the fetus may become a 'born again' patient" (24). For a critique of the rhetoric of "fetology," see Elaine Hoffman Baruch, Amadeo F. d'Adamo Jr., and Joni Seager, eds., *Embryos: Ethics and Women's Rights* (New York: Haworth, 1988), particularly the essay "Reproductive Technology and the Commodification of Life" (95–100), which analyzes the rhetoric of the fetus as a "product" subject to quality control.

83. Harrison, Golbus, and Filly, eds., *Unborn Patient.*

84. See Mark Seltzer's discussion of the dialectical relationship between what he terms the "privilege of relative disembodiment that defines the citizen of liberal market society ('abstract human personhood')" and the "making-conspicuous of the body in market society," in *Bodies and Machines* (New York: Routledge, 1992), 63.

85. On "mother," "motherhood," and reproductive discourse, see Valerie Hartouni, "Containing Women: Reproductive Discourse in the 1980s," in *Technoculture*, ed. Constance Penley and Andrew Ross (Minneapolis: University of Minnesota Press, 1991), 27–56.

86. Petchesky, "Foetal Images," 61–62. See also Barbara Katz Rothman, "The fetus in utero has become a metaphor for 'man' in space, floating free, attached only by the umbilical cord to the spaceship. But where is the mother in that metaphor?"; quoted in Petchesky, *Abortion and Woman's Choice*, xiv. For other recent examples of this argument, see Stabile, "Shooting the Mother"; and Adams, *Reproducing the Womb.* See also Marcus, "Placing *Rosemary's Baby*," for the related, if seemingly paradoxical, view of the woman as enemy of the fetus.

87. Richard Lewontin, quoted in Alfred I. Tauber, ed., *Organism and the Origins of Self* (Dordrecht and Boston: Kluwer, 1991), xvi.

88. On early anatomy and its relation to artistic representation, see William Anderson, "An Outline of the History of Art in Its Relation to Medical Science," *St. Thomas's Hospital Reports* 15 (1886): 151–81; Premuda, *Storia dell'iconografia anatomica*; William Schupbach, *The Paradox of Rembrandt's Anatomy of Dr. Tulp* (London: Wellcome Institute for the History of Medicine, 1982); essays in *Representations* 17 (1987), especially Glenn Harcourt, "Andreas Vesalius and the Anatomy of Antique Sculpture," 28–61; and Jonathan Sawday, "The Fate of Marsyas: Dissecting the Renaissance Body," in *Renaissance Bodies: The Human Figure in English Culture, 1540–1660*, ed. Lucy Gent and Nigel Llewellyn (London: Reaktion, 1990), 111–35.

89. See Gregor Martin Lechner, *Maria Gravida. Zum Schwangerschaftsmotiv in der bildenden Kunst* (Munich: Schnell-Steiner, 1981).

90. In his discussion of disembodiment and the citizen of liberal market society, Mark Seltzer points out the many "more deeply embodied bodies (in consumer society: the female body, the racialized

body, the working body)" against which the privilege of abstract universal personhood can be measured (63–64).

91. See Ludmilla Jordanova's discussion of the anatomical venuses in *Sexual Visions* (Madison: University of Wisconsin Press, 1989). She observes that in wax anatomy there are no male figures with bodies completely covered with flesh or recumbent (44). There are semirecumbent male figures in the Florentine collection, but they are flayed.

92. Although examples also exist of full-scale male figures, typically they are *scorticati* or "flayed ones" that present either the skeleton, nervous system, or musculature. In Bologna's anatomical collection, there are—uncharacteristically—two fully clothed bodies, busts of the Manzolini, the husband-and-wife team who executed the Bolognese collection and who preside over their wax creations. Giovanni Manzolini (Fig. 34) is dressed somberly in black with a lace-trimmed cravat; he holds a scalpel in his right hand and is dissecting a human organ. Anna Morandi Manzolini (Fig. 35), her human hair carefully coiffed, is dressed colorfully in silk, satin and lace, her earrings dangling, her face painted, her generous bosom draped lavishly with pearls. Her hands, ornamented with matching quadruple-stranded pearl bracelets and a large ring, are poised in the act of dissecting a human brain. She stares fixedly into space from her glass case against the wall, a striking contrast to the slender and shapely disemboweled odalisque she crafted that occupies the center of the room (Figs. 74, 75). On the semiotics of the wax venuses, see also Ludmilla Jordanova, "La donna di cera," *Kos: Rivista di cultura e storia scienze mediche, naturali e umane* 1 (1984): 82–86.

93. On Vesalius's frontispiece, see William Ivins Jr., "What about the *Fabrica* of Vesalius?" in *Three Vesalian Essays to Accompany the "Icones Anatomicae of 1934"* (New York: Macmillan, 1952); Erwin Panofsky, "Artist, Scientist, Genius: Notes on the Renaissance-Dammering," in *The Renaissance: Six Essays*, ed. Wallace Ferguson (New York: Harper & Row, 1962); and Devon L. Hodges, *Renaissance Fictions of Anatomy* (Amherst: University of Massachusetts Press, 1985).

94. On early anatomy, see J. B. de C. M. Saunders and Charles D. O'Malley, eds., *The Illustrations from the Works of Andreas Vesalius of Brussels* (Cleveland: World, 1950); Charles J. Singer, *A Short History of Anatomy from the Greeks to Harvey* (New York: Dover, 1957); and essays

in *Representations* 17 (1987), particularly Luke Wilson, "William Harvey's *Prelectiones*: The Performance of the Body in the Renaissance Theatre of Anatomy," 62–95.

95. On the pornographic aspect of the midwifery and obstetric literature, see, for example, the prologue to the popular and widely translated *De partu hominis*, paradoxically translated into English as *The Birth of Man-kinde; Otherwise Named the Woman's Booke*; the author alludes to objections that men read such books in order to know "the secrets and privities of women, and that every boy and knave had of these books reading them as openly as the tales of Robin Hood" (B8ᵛ). See also Jordanova, "La donna di cera."

96. David Freedberg, *The Power of Images: Studies in the History and Theory of Response* (Chicago: University of Chicago Press, 1989), 201. For an interesting consideration of the meaning of the anatomical classification of body parts, see Roy F. Ellen, "Anatomical Classification and the Semiotics of the Body," in *The Anthropology of the Body*, ed. John Blacking (New York: Academic Press, 1977), 343–73.

97. Robert Boyle, *Works* (London, 1772), 5:236. On the medical debates between mechanists and anti-mechanists in the late seventeenth and eighteenth centuries, see Sergio Moravia, "From 'Homme machine' to 'Homme sensible,'" *Journal of the History of Ideas* 39 (1978): 45–60. Ernst Cassirer, in *The Philosophy of the Enlightenment* (Princeton: Princeton University Press, 1951), claims the mechanist model "is an isolated phenomenon of no characteristic significance" (55). See also Robert E. Schofield, *Mechanism and Materialism: British Natural Philosophy in the Age of Reason* (Princeton: Princeton University Press, 1970); John Yolton, *Thinking Matter: Materialism in Eighteenth-Century Britain* (Oxford: Blackwell, 1983); and Stafford, *Body Criticism*. On the technological/machine model in contemporary obstetrics, see Barbara Katz Rothman, *In Labor: Women and Power in the Birthplace* (New York: W. W. Norton, 1982); and Robbie E. Davis-Floyd, "The Technological Model of Birth," *Journal of American Folklore* 100 (1987): 479–95.

98. Foucault, *Birth of the Clinic*, 171; see also Thomas Southwood Smith, "The Uses of the Dead to the Living" (originally published in *Westminister Review* 2 [1824]: 59–97), Jeremy Bentham's will and his essay "Auto-icon; or, Farther Uses of the Dead to the Living," and Simon

Shaffer's discussion in "States of Mind: Enlightenment and Natural Philosophy," all in *The Languages of Psyche* (Berkeley and Los Angeles: University of California Press, 1990), 233–90; see particularly Shaffer's discussion of the public dissection of Bentham.

99. *New York Times*, November 29, 1994, C8; I am grateful to Michel-André Bossy for this reference.

100. On recent biological challenges to the "individual," see Tauber, ed., *Organism and the Origins of Self*, xvi; Evelyn Fox Keller, *Secrets of Life, Secrets of Death* (New York: Routledge, 1992), esp. chaps. 6–9; and Stephen J. Gould, "A Humongous Fungus Among Us," *Natural History*, July 1992, 10–16. On the philosophical challenges to individualism, see Gilbert Simondon, "The Genesis of the Individual," in *Incorporations*, ed. Jonathan Crary and Sanford Kwinter (New York: Zone, 1992), 296–319. On Enlightenment protocols of science, see Keel, "Politics of Health."

101. Tauber, ed., *Organism and the Origins of Self*, xvi.

102. Wilson, "William Harvey's *Prelectiones*," 62.

103. Hunter's book was published by the renowned Baskerville Press with *face-à-face* Latin and English text, and was clearly a collector's item rather than a pedagogical manual, as Hunter's elaborate justifications of the format demonstrate. See Ludmilla Jordanova's fine discussion in "Gender, Generation, and Science." On anatomical atlases more generally, see David Armstrong, *Political Anatomy of the Body: Medical Knowledge in Britain in the Twentieth Century* (New York: Cambridge University Press, 1983).

104. For a full discussion of Hunter's atlas and eighteenth-century obstetrics, see Jordanova, "Gender, Generation, and Science."

105. On the notion of the "detail," see Naomi Schor, *Reading in Detail* (New York: Routledge, 1987).

106. See Jordanova, "Gender, Generation, and Science," 390–95, where these two engravings are compared to different, though related, ends; and Adams, *Reproducing the Womb*, 128–37.

107. For a powerful psychoanalytic account of the problem of the image and sexual difference, see Rose, *Sexuality in the Field of Vision*, particularly her discussion of the relationship between viewer and scene as one of "fracture, partial identification, pleasure and distrust" (227).

108. F. Gary Cunningham, Paul C. MacDonald, Norman F. Grant, Kenneth J. Leveno, and Larry C. Gilstrap III, eds., *Williams Obstetrics*, 19th ed. (Norwalk, Conn.: Appleton & Lange, 1993; first published by J. Whitridge Williams, 1903), 1. Subsequent in-text page references are to this edition unless otherwise specified.

109. R. V. Short, "Breast Feeding," *Scientific American* 250 (1984): 35—41, quoted in *Williams*.

110. See Robert A. Hahn's interesting discussion of *Williams*, particularly the shift from representing birth as a pathology to emphasizing its social and psychological aspects, in "Divisions of Labor: Obstetrician, Woman, and Society in *Williams Obstetrics*, 1903—1985," *Medical Anthropology Quarterly* 1 (1987): 256—87. Hahn speculates that the frontispiece was changed because "later authors of the text found it inappropriate or offensive, or believed that their readers would find it so" (261).

111. Cunningham et al., *Williams Obstetrics*, 8.

112. See the discussion of Bryson above, pp. 42—44.

113. Although it is often claimed that the "newness" of new visual technologies is their safety and noninvasive character, which is frequently contrasted with dissection and surgical procedures, in fact questions continue to be raised even about the safety of ultrasound; fetoscopy is known to pose considerable risk for both fetus and pregnant woman. On the new visual technologies, medicine, and industry, see Stuart S. Blume, *Insight and Industry: On the Dynamics of Technological Change in Medicine* (Cambridge, Mass.: MIT Press, 1992).

114. Sonographers and other technologists of the new visual apparatuses must "unlearn" the interpretive strategies of analog media because mimetic models produce false readings and, consequently, false diagnoses. See H. Schams and J. Bretscher, eds., *Ultrasonographic Diagnosis in Obstetrics and Gynecology* (New York: Springer-Verlag, 1975). On artifactuality and "reading" ultrasound images, see Sandra L. Hagen-Ansert, *Textbook of Diagnostic Ultrasonography*, 3d ed. (St. Louis: C. V. Mosby, 1989), 57ff. Commentators sometimes unwittingly note the interpretive problems posed by ultrasonography, as in so-called placental black holes, discussed in Franco Borruto, Manfred Hansmann, and

Juri W. Wladimiroff, eds., *Fetal Ultrasonography: The Secret Prenatal Life* (Chichester, Eng.: John Wiley, 1982).

115. On the "career" of ultrasound, its development in the weapons industry, and the politics of its success, see Blume, *Insight and Industry*. On ultrasound in obstetrical practice, see Schams and Bretscher, eds., *Ultrasonographic Diagnosis*; Edward Yoxen, "Seeing with Sound: A Study of the Development of Medical Images," in *The Social Construction of Technological Systems: New Directions in the Sociology and History of Technology*, ed. W. Bijker, T. Hughes, and T. Pinch (Cambridge, Mass.: MIT Press, 1987); and Hagen-Ansert, *Textbook of Diagnostic Ultrasonography*, which is dedicated "To our own little sonic boomers." On medical imaging and the nonmimetic, see Kember, "Medical Diagnostic Imaging." On modern medicine and the technologies of distance, see François Dagognet, *Ecriture et iconographie* (Paris: Librairie Philosophique J. Vrin, 1973), 88ff.

116. Rotman, *Signifying Nothing*, 19.

117. Jonathan Crary, *Techniques of the Observer: On Vision and Modernity in the Nineteenth Century* (Cambridge, Mass.: MIT Press, 1991), 1.

118. See Richard Wright's fascinating essay "Computer Graphics as Allegorical Knowledge: Electronic Imagery in the Sciences," *Leonardo*, "Digital Image—Digital Cinema" supplemental issue, 1990, 65–73.

119. For a critical reading of these new techniques of visual medicine, see Kember, "Medical Diagnostic Imaging." On visual culture and interpretation, see William J. Mitchell, *The Reconfigured Eye: Visual Truth in the Post-Photographic Era* (Cambridge, Mass.: MIT Press, 1992).

120. Howard Sochurek, *Medicine's New Vision* (Easton, Pa.: Mack, 1988); the article appeared in *National Geographic*, January 1987, 2–41. Figures 99–101 are reproduced from the original magazine article; Fig. 100 is a slightly different shot from that reproduced in the book.

121. Ibid., 146–47.

122. William M. Ivins Jr., *On the Rationalization of Sight*, Metropolitan Museum of Art Papers 8 (New York: 1938). See Elkins, *Poetics of Perspective*, for a powerful counterargument to Ivins.

123. Petchesky, *Abortion and Woman's Choice*, argues that "the neoconservative Reagan administration and the Christian Right accelerated their use of television and video imaging to capture political discourse.

. . . the anti-abortion movement has made a conscious strategic shift from religious discourses and authorities to medicotechnical ones, in its effort to win over the courts, the legislatures, and popular hearts and minds" (264). On the crucial importance of rhetorical commentary in producing ultrasound images on behalf of a "pro-life" position, see Condit, *Decoding Abortion Rhetoric*, 85ff. See also Valerie Hartouni's recent discussion of the problematic use of new visual technologies in both *The Silent Scream* and the "pro-choice" *S'Alines's Solution* in "Fetal Exposures: Abortion Politics and the Optics of Allusion," *Camera Obscura* 29 (1992): 130–49. On maternal bonding and the new visual technologies, see Petchesky, "Foetal Images," 73ff.; Joseph C. Fletcher and Mark I. Evens, "Maternal Bonding in Early Fetal Ultrasound Examinations," *New England Journal of Medicine* 308 (1983): 392–93; Marshall H. Klaus and John H. Kennell, *Bonding: The Beginnings of Parent-Infant Attachment* (New York: NAL, 1983); Beverley Hyde, "An Interview Study of Pregnant Women's Attitudes to Ultrasound Scanning," *Social Science of Medicine* 22 (1986): 587–92; L. S. Milne and O. J. Rich, "Cognitive and Affective Aspects of the Response of Pregnant Women to Sonography," *Maternal-Child Nursing Journal* 10 (1981): 15–39; C. L. Kohn, A. Nelson, and S. Weiner, "Gravidas' Responses to Real-Time Ultrasound Fetal Image," *Journal of OBGYN Nurse* 9 (1980): 77–80. For excellent critiques of the scientific literature on "bonding," see Arney, *Power and the Profession of Obstetrics*, 155ff.; and Diane E. Eyer, *Mother-Infant Bonding: A Scientific Fiction* (New Haven, Conn.: Yale University Press, 1992). On the recent use of sonograms in popular advertising, particularly the now notorious Volvo ad, see Taylor, "Public Fetus and Family Car"; and Rowland, *Living Laboratories*. It should be noted that in order to get a picture of the whole fetus, the sonogram must be taken at between twelve and sixteen weeks; subsequently, only fetal parts can be visualized.

124. See Leo W. Buss, *Evolution of Individuality* (Princeton, N.J.: Princeton University Press, 1987), 20.

125. Consider Paula Treichler's burgeoning list of terms: "egg mother, birth mother, name mother, surrogate mother, terry-cloth mother, gene mother, biomother, biomom, adoptive mother, legal mother, foster mother, mother of rearing, property mother, lab mother, blood mother,

organ mother, tissue mother, nurturant mother, earth mother, and den mother" ("Feminism, Medicine, and the Meaning of Childbirth," 130). See, among others, Patricia Spallone, *Beyond Conception: The New Politics of Reproduction* (Granby, Mass.: Bergin & Garvey, 1989); Marilyn Strathern, "Enterprising Kinship: Consumer Choice and the New Reproductive Technologies," *Cambridge Anthropology* 14 (1990): 1–12; and Hartouni, "Containing Women." For a history of distributed maternity in the context of race, see Hortense Spillers, "Mama's Baby, Papa's Maybe: An American Grammar Book," *Diacritics* 17 (1987): 65–81. On Baby M, see Janice Doane and Devon Hodges, "Risky Business: Familial Ideology and the Case of Baby M," *differences* 1 (1989): 67–82.

Epilogue

1. Andreas Huyssen cites filmmaker Alexander Kluge's virtually identical phrase in *Twilight Memories: Marking Time in a Culture of Amnesia* (New York: Routledge, 1995), 26.

2. Latour, "Visualization and Cognition," 3, 32.

3. Theodor W. Adorno, "Resignation," *Telos* 35 (1978): 166.

4. Theodor W. Adorno, *Negative Dialectics*, trans. E. B. Ashton (New York: Seabury Press, 1973), 159.

SOURCES AND CREDITS

Ornament. Cherubim holding obstetrical instruments. Relief. Joseph J. Plenck, *Anfangsgrunde der Geburtshulfe* (Vienna, 1774; earliest ed., 1768), title page.

1. Anti-abortion demonstrators, Washington, D.C. *New York Times*, January 23, 1992, A18. Angel Franco, NYT Pictures.

2. Illinois billboard, late 1980s. Photograph by Bruce Railsback.

3. Fetus at fifteen weeks. Photograph by Lennart Nilsson. *Life* magazine, April 30, 1965, 55. © Time Warner, Inc.

4. Fetus at eighteen weeks. Photograph by Lennart Nilsson. *Life* magazine, April 30, 1965, front cover. © Time Warner, Inc.

5. Fetus at eighteen weeks. Photograph by Lennart Nilsson. Reproduced in Nilsson, *A Child Is Born* (New York: Dell, 1965), 116–17.

6. Fetus at eleven weeks (1963). Photograph by Lennart Nilsson. *The Camera* (New York: Time-Life Books, 1970), 52; reproduced in Nilsson, *A Child Is Born* (New York: Dell, 1976), 86–87.

7. Sperm. Photograph by Lennart Nilsson. *Life* magazine, April 30, 1965, 56; reproduced in Nilsson, *A Child Is Born* (1976), 42–43.

8. Fetus at three and a half weeks. Photograph by Lennart Nilsson. *Life* magazine, April 30, 1965, 57; reproduced in Nilsson, *A Child Is Born* (1976), 48.

9. Fetal model. *New York Magazine*, April 24, 1989, 50. Copyright © 1989 K-III Magazine Corporation. All rights reserved. Reprinted with the permission of *New York Magazine*.

10. Anti-abortion demonstrators, St. Paul, Minnesota. *New York Times*, July 25, 1993, E3. Angel Franco, NYT Pictures.

11. Billboard, Providence, Rhode Island, 1993. Photograph by William Shotwell.

12. Abortion opponents, Minneapolis. *New York Times*, July 25, 1993, E3. Edward Keating, NYT Pictures.

13. Abortion opponent. *Christianity Today*, July 12, 1985, 40.

14. Advertisement from *Christianity Today*, July 12, 1985, 41. Courtesy of World Vision.

15. Still, *2001: A Space Odyssey*, directed by Stanley Kubrick. © 1968 Metro-Goldwyn-Mayer, Inc. Photofest.

16. Muscio, Lat. 7056, ff. 88–89, 13th cent. Bibliothèque Nationale de France, Paris.

17. Muscio, MS. 3701-15, f. 28, 9th cent. Bibliothèque Royale Albert Ier Brussels.

18. Eucharius Rösslin, *The Birth of Man-kinde; Otherwise Named the Woman's Booke*, trans. Thomas Raynald (London, 1626), H7^r; earliest ed. 1513. National Library of Medicine, Bethesda, Md.

19. Rösslin, *The Birth of Man-kinde*, H7^v. National Library of Medicine.

20. Scipione (Girolamo) Mercurio, *La comare o riccoglitrice* (Milan, 1618; earliest ed., 1596), Bb1^r. By permission of the Folger Shakespeare Library, Washington, D.C.

21. Jacob Rueff, *The Expert Midwife* (London, 1637; earliest German ed., 1554), K2^v. National Library of Medicine.

22. Hieronymus Fabricius, *De formato foetu* (Venice, 1627; earliest ed., 1600). National Library of Medicine.

23. Hendrik van Deventer, *Operationes chirurgicae novum lumen exhibentes obstetricantibus* (Leiden, 1701), fig. 16. National Library of Medicine.

24. Francis Mauriceau, *The Diseases of Women with Child and in Child-bed* (London, 1683; earliest French ed., 1668). By permission of the Folger Shakespeare Library.

25. Justine Dittrich Siegemund, *Spiegel der vroed-vrouwen* (Amsterdam, 1691), pl. 2. National Library of Medicine.

26. Van Deventer, *Operationes chirurgicae* (1701), title page. National Library of Medicine.

27. Lorenz Heister, *Institutiones chirurgicae* (Amsterdam, 1740; earliest German ed., 1718; Latin 1739), tab. 33. Yale University Harvey Cushing/John Hay Whitney Medical Library, Historical Collection.

28. Severin Pineau, *Opusculum physiologum et anatomicum* (Paris, 1597), Kiii^r. By permission of the Folger Shakespeare Library.

29. "An Human Ovum, about the third month," 22 December 1783; in Thomas Denman, *A Collection of Engravings Tending to Illustrate the Generation and Parturition of Animals, and of the Human Species* (London, 1787). Courtesy of the College of Physicians of Philadelphia.

30. "Three Human Abortions 23 February 1787"; in Denman, *A Collection of Engravings*. Courtesy of the College of Physicians of Philadelphia.

31. Rendering by Leonardo da Vinci (after 1487). © Her Majesty Queen Elizabeth II. With permission from Royal Collection Enterprises, Ltd.

32. Giotto, detail from *The Betrayal of Christ* (ca. 1304–13). Scrovegni Chapel, Padua. Alinari/Art Resource, New York.

33. Giovanni Antonio Galli (1708–82). Reproduced from *Ars obstetricia bononiensis* (Bologna: CLUEB, 1988), frontispiece. Courtesy of the University of Bologna.

34. Giovanni Manzolini (1700–1755). Reproduced from *Le cere anatomiche bolognesi del Settecento*, ed. Maurizio Armaroli (Bologna: CLUEB, 1981), 33. Courtesy of the University of Bologna.

35. Anna Morandi Manzolini (1716–74). Reproduced from Armaroli, ed., *Le cere anatomiche*, 35. Courtesy of the University of Bologna.

36. S. W. Fores, *Man-Midwifery Dissected* (London, 1793), frontispiece. National Library of Medicine.

37. From Ovid, *Trasformazioni*, trans. Lodovico Dolce (Venice, 1553), 197. Courtesy of the Boston University Libraries.

38. Twins (wax). Reproduced from *I Materiali dell'Istituto dell'Scienze* (Bologna, 1979), 244. Courtesy of the University of Bologna.

39. Pregnant uterus at monthly stages. Reproduced from *Ars obstetricia*, 73. Courtesy of the University of Bologna.

40. Pregnant uterus at seven months, in opened torso. Reproduced from *Ars obstetricia*, 73. Courtesy of the University of Bologna.

41. Birth sac and placenta. Reproduced from *Ars obstetricia*, 77. Courtesy of the University of Bologna.

42. Removal of the afterbirth. Reproduced from *Ars obstetricia*, 81. Courtesy of the University of Bologna.

43. Removal of the afterbirth. Reproduced from *Ars obstetricia*, 81. Courtesy of the University of Bologna.

44. Pregnant uterus lacerated during attempted delivery. Reproduced from *Ars obstetricia*, 81. Courtesy of the University of Bologna.

45. Difficult presentation. Reproduced from *Ars obstetricia*, 84. Courtesy of the University of Bologna.

46. Difficult presentation. Reproduced from *Ars obstetricia*, 87. Courtesy of the University of Bologna.

47. Prolapsed uterus. Reproduced from *Ars obstetricia*, 94. Courtesy of the University of Bologna.

48. Prolapsed uterus. Reproduced from *Ars obstetricia*, 94. Courtesy of the University of Bologna.

49. Uterine model in wire. Reproduced from *Ars obstetricia*, 33. Courtesy of the University of Bologna.

50. Uterine model in crystal. Reproduced from *I Materiali dell'Istituto delle Scienze* (Bologna: CLUEB, 1979), 245. Courtesy of the University of Bologna.

51. Monstrous birth. Reproduced from *Ars obstetricia*, 43. Courtesy of the University of Bologna.

52. Monstrous birth. Reproduced from *Ars obstetricia*, 100. Courtesy of the University of Bologna.

53. Speculum. Reproduced from *Ars obstetricia*, 59. Courtesy of the University of Bologna.

54. Obstetrical hook. Reproduced from *Ars obstetricia*, 53. Courtesy of the University of Bologna.

55. *Tirateste*, or head pullers. Reproduced from *Ars obstetricia*, 53. Courtesy of the University of Bologna.

56. Forceps. Reproduced from *Ars obstetricia*, 51. Courtesy of the University of Bologna.

57. Fetal presentation. Reproduced from *Ars obstetricia*, 35. Courtesy of the University of Bologna.

58. "Anatomie," plate XXII, *L'Encyclopédie* (1762). Brown University Library, Providence, R.I.

59. MS. 1122, f. 348^v, Bibliotheca Albertina. Courtesy of the University of Leipzig.

60. Andreas Vesalius, *De humani corporis fabrica* (Basel, 1543), 15^v. Brown University Library.

61. Petro Berrettini, *Tabula anatomicae* (Rome, 1741), tab. XXVII. Yale University, Harvey Cushing/John Hay Whitney Medical Library, Historical Collection.

62. From Cosme Viardel, *Observations sur la pratique des acouchemens naturels, contre nature et monstreux* (Paris, 1673), facing p. 12. National Library of Medicine.

63. Charles Estienne, *De dissectione partium corporis humani* (Paris, 1545). Sii^r. National Library of Medicine.

64. Estienne, *De dissectione partium corporis humani*, Siiii^v. National Library of Medicine.

65. Scipione (Girolamo) Mercurio, *La comare o riccoglitrice* (Venice, 1601), B2^r. National Library of Medicine.

66. From André Du Laurens, *L'anatomie universelle* (Paris, 1731), 45. National Library of Medicine.

67. Jourdain Guibelet, *Trois discours philosophiques* (Evereux, 1603), fig. II. National Library of Medicine.

68. Visitation of Mary and Elizabeth. Schotten Altarpiece (1390). Foto Marburg/Art Resource, New York.

69. Mary (wood, 1720–30). Copyright Bayerisches Nationalmuseum, Munich.

70. Adriaan Spieghel, *De formato foetu* (Padova, 1626), tab. IIII. National Library of Medicine.

71. Thomas Bartholin, *Bartholin's Anatomy* (London, 1668), Bb1^r. National Library of Medicine.

72. Hendrik van Roonhuyze, *Heel-konstige Aenmerckingen* (Amsterdam, 1663), frontispiece. National Library of Medicine.

73. J. B. Jacobs, *Ecole pratique des accouchemens* (Brussels, 1793), pl. 2. Bibliothèque de l'Académie Nationale de Médecine, Paris.

74. Anatomical model. Museo anatomico, Bologna. Photograph courtesy of Giovanna Ceserani.

75. Anatomical model. Museo anatomico, Bologna. Photograph courtesy of Giovanna Ceserani.

76. Anatomical model. Specola, Florence. Photograph by Liberto Perugi. Reproduced from *Le cere anatomiche della specola*, ed. Benedetto Lanza, Maria Luisa Azzaroli Puccetti, Marta Poggesi, and Antonio Martelli (Florence, 1979), 61.

77. Anatomical model. Specola, Florence. Photograph by Liberto Perugi. Reproduced from Lanza et al., eds., *Le cere*, 227.

78. Miniature anatomical model. Bibliothèque de l'Académie Nationale de Médecine, Paris.

79. Andreas Vesalius, *De humani corporis fabrica* (1543), frontispiece. Brown University Library.

80. Etruscan ex-voto, uterus, 3d cent. By permission of the Trustees of the British Museum, London. Copyright British Museum.

81. Anatomical detail. Museo anatomico, Bologna. Reproduced from Armaroli, ed., *Le cere anatomiche*, 63. Courtesy of the University of Bologna.

82. Anatomical detail. Museo anatomico, Bologna. Reproduced from Armaroli, ed., *Le cere anatomiche*, 58. Courtesy of the University of Bologna.

83. Anatomical detail. Museo anatomico, Bologna. Reproduced from Armaroli, ed., *Le cere anatomiche*, 55. Courtesy of the University of Bologna.

84. Anatomical detail. Museo anatomico, Bologna. Reproduced from Armaroli, ed., *Le cere anatomiche*, 93. Courtesy of the University of Bologna.

85. Anatomical detail. Museo anatomico, Bologna. Reproduced from Armaroli, ed., *Le cere anatomiche*, 88. Courtesy of the University of Bologna.

86. Sexless multi-torso. Anatomical catalog. Photo courtesy of Denoyer-Geppert Science Co., Chicago.

87. Pregnancy insert. Anatomical catalog. Photo courtesy of Denoyer-Geppert Science Co., Chicago.

88. Pregnant uterus set. Anatomical catalog. Photo courtesy of Ward's Natural Science Establishment, Inc., Rochester, N.Y.

89. Jan van Rymsdyk (artist), in William Hunter, *The Anatomy of the Human Gravid Uterus* (1774), pl. VI. Brown University Library.

90. Jan van Rymsdyk (artist), in William Smellie, *A Sett of Anatomical Tables* (1754), pl. X. Courtesy of the College of Physicians of Philadelphia.

91. From William Cowper, *The Anatomy of Human Bodies* (1698). Brown University Library.

92. Charles N. Jenty, *Demonstratio uteri praegnantis mulieris* (1759), fig. 4. National Library of Medicine.

93. Jacques Gautier Dagoty, *Myologie complette en couleur et grandeur naturelle* (1746). Brown University Library.

94. J. Whitridge Williams, *Obstetrics: A Textbook for the Use of Students and Practitioners*, 3d ed. (New York: Appleton, 1912; earliest ed. 1903), frontispiece. Brown University Library.

95. F. Gary Cunningham et al., eds., *Williams Obstetrics*, 19th ed. (Norwalk, Conn.: Appleton & Lange, 1993), 380.

96. Cunningham et al., *Williams Obstetrics*, 382.

97. Cunningham et al., *Williams Obstetrics*, 274.

98. Cunningham et al., *Williams Obstetrics*, 276.

99. Throat tumor. *National Geographic*, January 1987, 26. Courtesy Howard Sochurek/National Geographic Image Collection.

100. Joseph Ward, *National Geographic*, January 1987, 26. Courtesy Howard Sochurek/National Geographic Image Collection.

101. Yawning fetus. *National Geographic*, January 1987, 25. Courtesy Howard Sochurek/National Geographic Image Collection.

102. Fetus at 12–16 weeks. Volvo advertisement, 1990. Courtesy of Volvo.

103. Human fetus X-ray, ca. 1980. Science Source/Photo Researchers, Inc.

INDEX

In this index an "f" after a number indicates a separate reference on the next page, and an "ff" indicates separate references on the next two pages. A continuous discussion over two or more pages is indicated by a span of page numbers, e.g., "57–58." *Passim* is used for a cluster of references in close but not continuous sequence.

Library of Congress Cataloging-in-Publication Data

Newman, K.
    Fetal positions : individualism, science, visuality / Karen
Newman.
        p.   cm. — (Writing science)
    Includes bibliographical references and index.
    ISBN 0-8047-2647-7 (cloth : alk. paper). — ISBN 0-8047-2648-5
(pbk. : alk. paper)
    1. Fetus—Social aspects.   2. Fetal monitoring—Europe—History.
3. Visual perception—Europe—History.   4. Image processing—Europe.
5. Human reproduction—Europe.   6. Sex role—Europe.   7. Body,
Human—Social aspects.   I. Title.   II. Series.
    RG600.N49   1996
    618.2'0094—dc20   95-41182
                                                                  CIP

Original printing 1996
Last figure below indicates year of this printing:

05   04   03   02   01   00   99   98   97   96

♾ This book is printed on acid-free, recycled paper.